The Architect and the Scaffold

The Architect and the Scaffold

Evolution and Education in South Africa

Edited by
Wilmot James
Lynne Wilson

Human Sciences Research Council

New Africa Education

Compiled by the Social Cohesion and Integration Research Programme, Human Sciences Research Council (Executive Director: Wilmot James)

Editors: Wilmot James and Lynne Wilson

Published by the Human Sciences Research Council Publishers
Private Bag X9182, Cape Town, 8000, South Africa

and

New Africa Education, an imprint of New Africa Books (Pty) Ltd
99 Garfield Road, Claremont, 7700, South Africa

First published 2002

ISBN 0-7969-2003-6

Produced by comPress

Printed by Creda Communications

Distributed in South Africa by Blue Weaver Marketing and Distribution, P.O. Box 30370, Tokai, Cape Town, South Africa, 7966. Tel/Fax: (021) 701-7302, email: blueweav@mweb.co.za

Contents

Preface

The Colloquium on Science and Evolution in the Fullness of Life held at Spier Estate's Conference Centre on 8 June 2002 predictably generated a great deal of debate in our newspapers. This is good because the first step in advancing knowledge is to talk about the issues, especially the controversial ones.

And evolution is probably the most controversial of all, for it cuts deep into our common-sense understanding of who we are, where we come from and where we are heading as an advanced species populating and destroying, and as we create and live in this world. Evolution is about understanding human nature; no wonder, therefore, that it generates so much 'heat.'

But the public debate illustrates also the challenges of public understanding of science, human biology and evolutionary theory. Indeed, there is no common vocabulary to discuss the issue rationally; for example, scientists use the term 'theory' to refer to a confirmed and coherently organised body of tested fact, while your average citizens think theory is mere speculation.

This book intends to go beyond the limits of the public discourse to give more depth to the challenges evolution poses to education in South Africa. It seeks not only to fill the gaps in public knowledge but also to provide a frame of reference by which we can better understand the facts of everyday life. It is therefore part of the architecture and scaffolding of knowledge about ourselves, the nature of our family lives, religion and spirituality, art and culture, sexuality, and even sports and politics.

We are grateful to all contributors to this volume, all of whom save David Chidester, whose chapter was specially commissioned, gave presentations at the Colloquium on Science and Evolution in the Fullness of Life. Gratitude also goes to Shell and the Swiss Agency for Development and Co-operation (SDC) for making the Colloquium possible.

The Architect and the Scaffold is dedicated to the late Stephen Jay Gould, who made an extraordinary contribution to the public understanding of human biology and evolution.

WILMOT JAMES
Executive Director
Social Cohesion and Integration Research Programme
Human Sciences Research Council
August 2002

A Tribute to Stephen Jay Gould

Pippa Skotnes

My first encounter with Stephen Jay Gould was in the late 1970s when I was a second year Archaeology student at the University of Cape Town. It was an encounter characterised by a wonderful academic generosity and a most extraordinary curiosity that seemed prepared to find enlightenment in the most unlikely of places. We had just had a series of lectures on the theories of gradualism and punctuated equilibrium. I recall being highly excited thereafter, and was delighted to identify, in an essay we had to write on the merits of each theory, what I saw as a few flaws in Gould's thinking. I wrote to tell him this. To my surprise (though it is even more surprising to me now, when I think back on this) he wrote back, concerned to know exactly what my problems were and asking me to keep in touch. I received a couple of postcards a while after when he was travelling in Italy; he was still worrying about the flaws, though by this stage I had realised that, as a 20-year-old Fine Art student with a passing interest in evolution, I wasn't going to able to hold up my end of the debate, and I steered our correspondence into other areas.

Sometime later Steve contributed to a book I published on the southern San. He loved the idea that the energies of a scientist could lie side-by-side those of an artist on the pages of a book. He also provided me with a number of reflections on various topics, particularly when I was working on a project that examined, in part, aspects of the history of scientific racism in South Africa. His *Mismeasure of Man* had been a great influence on my own thinking about the history of intelligence-testing and racism, as it has for many others, not least of all in South Africa. It was guided by deep compassion and a true understanding of the destructive effects of negative stereotyping. Similarly, his work on the iconography of evolution has been crucial in helping me understand just how stereotypes are generated and systems of racism are justified. This work of his has been central in providing ways of explaining human diversity and underlining the dangers of equating evolution with progress. Steve believed that evolution should be taught in every classroom (uncontested by creationism or other theories of origin) – an imperative made even more important in South Africa, I believe, given our own troubled past and the devastating effects of a belief in racial privilege. His legacy in this area is profound.

Steve Gould had undoubtedly the most extraordinary and active mind I have encountered. When I first met him, I recall standing, a little overwhelmed, in his huge office – surrounded by hundreds of cabinets with overflowing drawers and rows and rows of shelves of books. His desk was bare, but for a photograph of himself with the pope, and a 1920 Smith Corona typewriter on which he banged out his essays, word perfect and without correction or second draft. He talked incessantly and often at great speed. He had perfect recall of almost everything he read and heard, and it was, at times, a frustration for him that his speech could not keep up with the stampede of thoughts and ideas that moved through his head. At that time he was excited to show me a piece he'd just completed on Leonardo's Leicester Codex. 'I've never spent so much time on an article,' he announced. 'How long was that, Stephen?' I asked. 'Oh all, of two weeks,' he replied. 'Two weeks!' I gasped. He looked apologetic – 'Well,' he said defensively, 'it's not easy to read medieval Italian in a mirror!'

This year, 2002, provided a number of moments of closure for Steve. His great tome on the structure of evolutionary theory was published in March. When I saw him last it was almost complete, and he was looking forward to the flurry of comment he knew it would provoke. He had completed his run of essays for Natural History and his final compilation was published shortly before his death. And there was an end of sorts to the story of Saartjie Baartman, whose skeleton was returned to South Africa in July 2002 after much political wrangling. Steve had come upon her remains in the Musee de l'Homme in Paris and his essay about her had been responsible for resurrecting her from historical obscurity and providing South Africans with an icon of colonial humiliation.

Nevertheless his death was as untimely as it was unexpected. There were new books he was working on, and an almost implausibly diverse range of anecdotes still to be drawn into discussions about science, evolution and history. He was an extraordinary, often provocative, scientist, but it was this range, this ability to capture the richness of human experience and bring science and evolution to bear upon it, that made him also an extraordinary human being and his death such a loss to scholars and public alike.

Introduction

We live in exciting times. The 20th century saw humankind move into an age of technology. We explored and mapped the surface of our earth, the depths of our oceans and our immediate planetary neighbours. But it was not until late in the century that we began what is possibly the most challenging and relevant exploration of all: the mapping of our own genome.

Our exploration of the building blocks of life has naturally led us to question other aspects of our being: Why are we here? How did we get here? In whose hands are we going to entrust the knowledge of the most intimate details of ourselves?

Other questions also arose: When Africa is widely acknowledged as the cradle of humankind, why was our continent marginalised in the genome research? What were the ethical, moral and legal implications of this research for a largely uninformed population?

The Africa Human Genome Initiative was founded to address these concerns.

The human genome is actually more like a library than a map, with a myriad books ready to be opened and read by generations of scientists. The more we learn about the human genome, the more we discover there is to explore. And the wider we prise open this Pandora's box, the more controversy and acrimony – especially in the fields of science, evolution and religion – we release in the process.

The story of genetics has been closely linked to education and religion since it began.

Gregor Mendel, the 'monk in the garden', as his biographer Robin Marantz Henig describes him in the title of her engaging book, was an abbot and a teacher, as well as a gardener and a fervent naturalist, whose curiosity and keen observation of pea plants led to his discovery of the basic laws of heredity and the beginnings of the modern science of genetics.

Mendel's first paper was published in 1866, just seven years after Charles Darwin's seminal *Origin of Species,*[1] and although it did not provoke the same fervour from its adherents or criticism from its detractors, it was the first step on the road that has led, today, to the unravelling of the web of life that is the human genome.

Mendel read Darwin's theories with great interest, but his own work was largely unrecognised in his lifetime, with the result that Darwin was never aware of Mendel or his triumphant scientific experimentation, although Darwin knew that the lack of an explanation for heredity left a gap in his theory of natural selection.

Who knows what advances were lost because of the gap in Darwin's knowledge? And who knows what damage we could do to our own children and the future scientific progress of our nation by not providing full knowledge of some of the most relevant research of our time?

The question of what is taught in South African schools is not a simple one. Under the apartheid government and its Christian National Education policy, evolution was given scant recognition in the curriculum. Add to this the fact that there was a policy of exclusion from science education for black learners, and the scale of the problem of reforming our children's science education becomes apparent.

Jeffrey Lever sparked a debate in his paper *Science, Evolution and Schooling in South Africa,*[2] republished in this book. His comments issued a challenge to all involved in the educational sector and demanded a response to the question of whether we are giving the proper attention to evolutionary theory in our schools.

As Bernard Lategan asks later in this book, 'How should we reform our school system and school curricula in the new South Africa if we want to avoid the pitfalls of the past? Can we find a

1 Darwin, C. 1956 [1859], *The Origin of Species*. J.M. Dent: London.

2 Lever, J. 2002, *Science, Evolution and Schooling in South Africa*, Africa Human Genome Initiative: Cape Town.

more constructive approach to the teaching of science and of religion? Surely the main thrust of Lever's argument is that the South African school system – for whatever reason – has failed to prepare students to deal with and to benefit from the insights of contemporary science, specifically as far as the origin and evolution of life is concerned.'

As the Human Genome Project opens the door to gene patenting and bioethics, we are fully aware that an uneducated population is a potentially exploited one. We cannot afford to sit back and allow the genetic revolution to pass us by; we must tackle the issues that arise from the research and we must be prepared to face and, if necessary, allay the concerns of all the constituencies that are affected by them.

In his paper commenting on and responding to Lever's *Science, Evolution and Schooling in South Africa,*[3] Wieland Gevers postulates that the question should not be whether evolution is taught in our schools, but rather what aspects of it are taught.

He argues that, even with the 'gap between the brilliance of scholars who can reduce the huge complexity of their fields of enquiry to a simple, understandable level in books sold in up-market bookshops, and the reality of thousands of classrooms filled with restless teenagers conducting a personal "arms race" (not the evolutionary kind) against their teachers, the syllabus and textbook learning in general', evolution is a subject that cannot be ignored.

He argues further that our children's education should not stop there; evolution should be used as a springboard to tackle issues such as HIV and other 'big issues of human life on earth'.

He throws down the gauntlet to pedagogical thinkers with his contention that the major trends in evolutionary thought can be accessible to the average adolescent, if presented in a relevant and accessible manner.

Taking the concept a step further, Linda Chisholm assesses our school textbooks in the light of the development of the curriculum. The issue of what is taught, and how, is uppermost in her mind and the minds of the authors of the Revised National Curriculum

3 Lever, J. 2002, *Science, Evolution and Schooling in South Africa.*

Statement. She and her colleagues are perhaps more aware than most of the public response to evolution, and the challenges posed by the 'extremely unequal resources and conditions of schooling in South Africa'. Her paper is a response to these questions; it lucidly outlines the procedure of curriculum development, and the challenging and unenviable task of trying to please all interest groups. Ultimately, she says, 'the curriculum was shaped by a multiplicity of new, diffuse social forces, voices and educational philosophies, much less visible and loud than the evangelicals, but also far more powerful in the new South Africa'.

Naledi Pandor continues the story. She calls for a curriculum that does more than focus on 'previously excluded areas of study and exciting new stories'. While she supports the call for evolution to be taught in our schools, she is not convinced by Jeffrey Lever's assertion that 'excluding evolution damages and limits learners now' and that 'learners in the past have been damaged due to the exclusion of evolution from the curriculum under apartheid'. In fact, she warns, Lever's position may do little more than 'succeed in reviving the largely silent Calvinist lobby of yesteryear, and provide it with a large whipping stick for stirring up religious fervour and furore in our schools'.

Rather than concentrating only on evolution, she says, we must ensure that moral issues and norms are taught in tandem with the scientific theories and advances which must form an integral part of our children's education. Both the useful and malign influences of science need to be clear.

Pandor also inserts a caveat into the debate with her assertion that Darwinian evolution has been used to 'give support to repugnant racial theories and racist movements'.

Catherine Odora Hoppers takes up the issue with her passionate plea for the education of our children to go further than the debate between 'the evolutionist goats and the creationist sheep'.

Taking her position as a 'radical witness and wounded healer', Odora Hoppers suggests that the battle lines are drawn not just between evolutionists and creationists. She inserts another front by adding imperialism and colonialism and the battle for the hearts and minds of indigenous people. Indigenous knowledge and traditional values have been eroded and belittled by the 'mirror of western identity'. If we are changing our curriculum, she argues, we should ensure that all the paradigms of our society are equally represented.

Science, she also argues, is alienating in that it turns the individual into an observer, a mere recorder of events. In addition, it panders to the male obsessions of domination and control, while leaving Africa a continent bereft of her virginity and robbed of her self-worth.

She issues a challenge to scientists and curriculum developers alike to 'help science recover its sense of dwelling, its sense of caring, crying, laughter and joy'.

Abdulkader Tayob also bemoans the missing links in Lever's paper, albeit from a Muslim perspective. While the fundamentalist Islamic view would find itself a bedfellow of the fundamentalist Christian perspective in its rejection of evolution, Tayob furthers the argument with his counterpoint Islamification of science: 'According to this intellectual trend, Muslim scientists are called upon to restructure modern sciences, social and natural, on the basis of Islamic values' – many private Islamic schools are embracing this concept.

Echoing the views of Pandor, Muslims call for a religio-cultural framework for science to provide a moral basis with which to assess the clinical conclusions of scientific endeavour.

For Tayob, it is the gaps in Lever's paper that beg more questions than the paper itself: 'The discussion is entirely restricted to white educational experience of evolution and Darwin in South Africa. The Islamic responses cited … are the only revelation that South Africa has more than one educational tradition, and a minuscule one at that. And even then, Muslim responses are presented as a reflection of Christian fundamentalism.'

In addition, he takes Lever to task for his dogmatic assertion that evolution 'can unlock the secrets of the universe and of the human condition itself'. In his own experience (we're sure shared by many, of whatever religious persuasion), evolution inspired an understanding of biology but in no way shook his hold on the Qu'ran as a central pillar of his life.

His paper serves as a warning against a monoculture of thought. As he says, 'The pluralistic history and experience of South Africa is much too precious and rich to be obliterated with one framework.'

Continuing the religious debate, Denis Davis enters the fray from what he describes as a 'somewhat dissident Jewish perspective'.

He takes us back to the beginning, building a case for understanding evolution within the first chapters of the Book of Genesis. For Davis, the story of our evolution is hidden in the Hebrew words chosen by the writer of Genesis. He explains the tradition of Jewish mysticism which 'viewed the world as continuously evolving towards the goal of ultimate perfection'.

He issues a dire warning against the kind of fundamentalism that asserts that 'there is only one text, there is only one interpretation, there is only one answer', saying that it is this type of thinking that threatens human freedom, democratic society and the values prefigured in our constitution.

In spite of Davis's appeal, the letters pages of our daily newspapers at the time of the Colloquium indicated that the schism between the evolutionists and the creationists seems inevitable and unbridgeable. In his paper, David Chidester attempts to open the debate by calling for the protagonists not to accept or reject the proposition but rather to renegotiate the terms in which that proposition is formulated.

As an example he draws on the history of Christianity in Africa, where he says Africans accepted or rejected the appeal of the Christian gospel and also renegotiated its religious terms. He suggests that there is room for renegotiation of the terms under which evolution is taught, but also observes rather wryly that 'knowledge is power' and as such 'is inevitably mixed up with political interests'.

So to the curriculum itself. Dev Isaac outlines the process of developing a curriculum within the critical and developmental outcomes and assessment criteria of the National Curriculum Statement. Perhaps the most vital core of this whole debate can be summed up in this statement: 'The learner will be able to act confidently on curiosity about natural phenomena, and to investigate relationships and solve problems in scientific, technological and environmental contexts.'

While Isaac tells us what should be done, Fathima Dada and Colleen Dawson tell us how these aims can be translated into actual lessons in the classrooms of our nation.

Dada tells us of the publishers' dilemma: How does one construct a text book which will remain fresh and relevant in a rapidly changing and ever-expanding scientific environment? Should creationism be taught as part of the religious education curriculum, with evolution being taught in the Science class? There is a hint of desperation in her tone as she is faced by Lever's claims that evolution is 'uncontroverted' and asks just 'who claims what is uncontroverted?'

She echoes the call of so many others in this book when she says, 'The notion that the school curriculum should be a tool for propagating particular theories is a dated one. We should rather use the opportunity to provide our children with the knowledge and skills to analyse a variety of views and arguments, encouraging them to engage in debate, analysis and the curiosity to seek more knowledge.'

How to nurture this creativity and curiosity is the particular interest of Colleen Dawson, whose paper gives flesh to the bare bones of curriculum and textbook. She takes us into the classrooms and playgrounds and provides a glimpse of education that challenges and stimulates young minds to reach their fullest potential. As we discover more about the human genome, the controversial subject of cloning has sparked much debate. After reading Dawson's paper, however, I'm sure there would be widescale support for cloning a Ms Dawson for each of our schools.

If the designer Edwin Schlossberg will allow us a little poetic licence with his statement on writing, perhaps the skill of curriculum development is 'to create a context in which other people can think'. When we can do this, we will truly be offering our children an education that lives up to its promise in the 'fullness of life'.

Thanks are due to The Swiss Agency for Development and Co-operation, and in particular Gerhard Pfister, without whose generous support this book would not have been published. We are also grateful to Shell for their support of the Colloquium at which the papers were presented.

Editors
Wilmot James
Lynne Wilson
Cape Town, July 2002

Science, Evolution and Schooling in South Africa

1

Science, Evolution and Schooling in South Africa

Jeffrey Lever

Introduction

> The recent record reinforces the lesson of the years since 1858: that Darwin's account of evolution remains intact. Hereditable variation and natural selection are indeed the agents that shaped the present richness of life on Earth. The new genetics does not challenge Darwinism but, on the contrary, is the means by which the details of the course of evolution will be unravelled from the sketchy fossil record and the growing accumulation of data about the genetic constitution of animals and plants.[1]

In 1952 the curators of the Transvaal Museum in Pretoria exhibited some of the findings of South Africa's small band of palaeontologists regarding the evolution of early humans in southern Africa. It was an appropriate time. After more than 20 years of imperial disdain towards the upstarts of colonial science, Raymond Dart and his former colleague Robert Broom at the University of the

1 Maddox, J. 1998, *What Remains to Be Discovered: Mapping the Secrets of the Universe, the Origins of Life, and the Future of the Human Race*, Macmillan: London.

Witwatersrand were vindicated. With none other than that doyen of British palaeontology, Sir Arthur Keith, in the lead, world science had at last placed its stamp of approval on the hominid status of Dart's famous 'Taungs Baby', *Australopithecus africanus*.[2] It was in Africa that the human lineage had first evolved, not Asia, as European and American scientists had previously believed.

Not everyone was pleased however. Representatives of the three Dutch Reformed Churches contacted the Museum's curators and made clear their strong objections to the exhibition. The view of the three Reformed Afrikaans Churches was that evolution was no more than a hypothesis, and a far-fetched one at that. More to the point, it was in conflict with the Bible, the early part of the Book of Genesis in particular. Even scientists themselves, wrote the editor of *Die Kerkbode*, (the official bulletin of the *Nederduitse Gereformeerde Kerk*, by far the largest of the three Churches), disagreed on evolution. The *dominees* made much of Robert Broom's own heterodox views on evolution.[3] 'We are grateful,' wrote *Die Kerkbode's* editor, 'that this matter has been raised once again by our leaders of the church and trust that it will not happen again, as so many times in the past, that the views of the Afrikaans Churches on this issue will be ignored'.[4]

This minor skirmish between the clerics and the scientists occurred in the shadow of much larger events. Government segregationists and Defiance Campaigners had bigger fish to fry during 1952. But the pressure on the Museum curators was very much part of a broader political tapestry, one being lowered over the country in the name of a brand of Protestant Christianity that commanded only a relatively small minority of the country's Christian believers. Among the many victims of Christian National Education, Darwin's theory of evolution by natural selection was but one, and in social terms a relatively minor one, offered up to the civil religion of ethnic nationalism during the

2 Dart, R. 1959, *Adventures with the Missing Link*, Hamish Hamilton: London, Chapter Seven.

3 Broom, like Arthur Russel Wallace, co-founder with Darwin of the theory of evolution by natural selection, considered that there had been some 'spiritual agency' at work in the evolution of humankind. See for example Wells, L.H., 1967, 'One Hundred Years: Robert Broom 30 November 1866 – 6 April 1951'. Robert Broom Memorial Lecture. *South African Journal of Science*, vol. 63, September, p.362.

4 *Die Kerkbode*, 1949–1952, 24 September 1952.

course of the 1950s. But it is one whose spectre is only now, very gingerly it would seem, being laid to rest more than 50 years later in the nation's schools.

Evolution and Science at the Inception of the 21st Century

Over the last 150 years modern science has assembled a remarkably coherent picture of our universe and humanity's place within it. We now can give evidence-based answers to the questions that humans have posed since the dawn of consciousness. What is this reality all around us? What is this starry firmament that shines down on us here when we care to look up? Where did it come from? How did life arise and with it the human being?[5] How indeed is there *anyone at all* out there looking into the night, and asking these questions?

These issues are of more than metaphysical interest. The scientific revolution of the last 200 years has proceeded hand-in-hand with technological advance. The culminating scientific achievement brought about by 20th century science, the unravelling of the human genome in mid-2000, could only take place due to the developments in computation that began a mere 50 years ago. Without the high-powered computers that derive from the intellectual breakthroughs in mathematics in the 1930s, the task of sequencing the 3.2 billion base pairs of our common human genome would have been unthinkable. The connection runs deeper. The binary maths of the computer programme is uncannily paralleled by the linear digital code of the DNA sequence. Thus basic science and technology advance in tandem. Our technology rests on our understanding of the scientific fundamentals that are derived from the overarching theories brought to fruition in the 20th century. The technology in turn deepens our theoretical progress in a continual feedback loop.

At the beginning of the 21st century, then, our scientific synthesis regarding many of the most salient features of our exter-

5 For a review of current notions on the origin of life, and indeed on the other major frontiers of contemporary science, see the book by the former editor of *Nature*. Maddox, J. 1998, *What Remains to be Discovered: Mapping the Secrets of the Universe, the Origins of Life,* and *the Future of the Human Race,* Macmillan: London.

nal reality hangs together as never before. There is a consistent narrative to be told about our cosmic beginnings, our middle passage through star and galactic formation, and the ultimate genesis of life on earth, whether in some 'warm little pond' as Darwin hypothesised, or elsewhere. The story is far from complete. Parts of it may be altogether mistaken. A new twist in one part may cause ripples that reconfigure even the fundamentals of the current picture – just as Max Planck's small step towards the idea of a quantum of energy was to upset the whole of a seemingly finished physical science at the beginning of the last century.

But let us not underestimate the achievements of the 20th century scientific endeavour. It has abolished once and for all the mechanistic determinism that Newtonian physics seemed to install in the universe. Materialism as a doctrine now rests on shaky foundations when we probe the inner recesses of matter. The universe has changed its character from one of infinite space existing in eternal time to a space-time bounded explosion of potential that brings back to mind the discredited metaphysics of G.W.F. Hegel. The debate over the precise implications of quantum mechanics for our ontology still rages. Here in the paradoxical interplay of wave and particle, in the mysterious non-local entanglement of far distant particles, can still be found some room for theological and metaphysical musings. Not that most scientists themselves are inclined to see in all this any reason for a new supernaturalism or deism.[6] Yet things are definitely more interesting than when a clock-work universe seemed the appropriate cosmic model.

Strangely it is the most accessible, least quantified theory of 19th century science that has best stood the test of time. Anyone can read *The Origin of Species* and get the gist of the message. Darwin's theory of evolution by way of natural selection has been elaborated, refined, and combed over for possible flaws, but it still stands stubbornly uncontroverted. Its power has seemed to dim several times over the past 140 years, but each time it has returned, more compelling than before in its stark logic.

6 Nobel Prize-winning physicist Steven Weinberg made the famous comment that still resonates in the debate: 'The more the universe seems comprehensible, the more it seems pointless.' (Weinberg, S. 1993, Dreams of a Final Theory. Hutchinson Radius: London.

Even the single most glaring gap in his theory, the mechanism for the transmission of inherited characteristics, about which Darwin cooked up various but totally wrong notions, was to strengthen his reasoning. (Ironically, Gregor Mendel had sent Darwin a copy of his ground-breaking paper on the pattern of inherited traits in peas, but it remained in the famous man's library, uncut and unread.) Inheritance was not the blending of parental characteristics – a process that would have been fatal to Darwin's theory – but particulate, or as we now say, by way of genes. In other areas where Darwin only speculated, as in the origin of humankind, and, more groundedly, on the evolutionary development of behaviour and culture, later scientists were to find his points of departure extraordinarily fruitful.

What Darwin immediately offered biology in the mid-19th century was a framework that was at once an explanatory *and* an ordering one. Through 'descent with modification', a family of self-replicating entities could hand down their inborn traits to offspring in an environment that offered only scarce energy. To survive and reproduce necessarily became a competition for this energy once the copying process had got firmly under way. Any minor modification that offered even the slightest advantage in the hunt to refuel and replicate would lead – probabilistically – to the increased reproduction of one copy over another, slightly differing one. So the process that we infer *must* have happened around 3.8 billion years ago got underway, and the logic of self-replication in a demanding environment led to the operation of natural selection. As Darwin concluded, 'from so simple a beginning endless forms most beautiful and most wonderful have been, and are being, evolved'.[7]

It is thus only by harnessing the basic theory that Darwin sketched in *The Origin of Species* that the following questions can be answered and the pillars of the discipline of biology put in place:

What *is* life? It is precisely the current (Darwinian) understanding of what makes something 'living' that enables us to grasp many otherwise inexplicable features of both the organic and the

7 This is the last sentence of *The Origin Of Species*, (Darwin, C. 1956 [1859], The Origin of Species, J.M. Dent: London.

inorganic world.[8] Life is complex self-replication under the pressure of natural selection.

Why do we find such a wealth of organic diversity? Biologists have put the number of species at around 10 million, though confessing it could be many times more. The diversity of life, the explanation-begging fact of difference, can only be understood by reference to the evolutionary past of around 3.8 billion years.

Why do we find unity in this diversity? It is clear even to the least biologically informed that living forms show a striking family resemblance. That fact was known thousands of years ago, to Aristotle among others. Darwin boldly argued that the phenomenon could be explained if we hypothesised that all living forms derived from one original ancestral being. This leap of scientific intuition – without any conclusive evidence at the time – has been resoundingly confirmed with the discovery in the 1950s of the common genetic mechanism programmed into the DNA in all organisms.

How can we bring order to this immense organic diversity and relate the various groupings in one consolidated scheme? The five kingdoms of life into which biologists classify all organisms constitute an evolutionary schema, beginning with the earliest and simplest to the later explosion of multi-cellular beings after the Cambrian Era around 530 million years ago. Again, this classification brings order and structure to the study of living beings, but its rationale is a thoroughgoing evolutionary one that is now being put on the new basis of comparative genomics through DNA analysis. The modern evolutionary synthesis is the frame within which the advances of the life sciences take place, making sense of the immense variety of seemingly unconnected fields.[9]

Nowhere is this fact more evident than in the revolution underway in the wake of the near-completion of the Human Genome Project. The very shape of our genomic heritage – bewilderingly complex, bafflingly anarchic – testifies to our nearly four billion

8 For a persuasive account of 'life' as replication, see Dawkins, R., (new edition), Oxford University Press: Oxford, 1989, *The Selfish Gene* and, also by the same author, 1982, *The extended phenotype, The gene as the unit of selection*, Oxford University Press: Oxford.

9 Here one could also repeat the often-cited remark by the leading 20th-century biologist, Theodore Dobzhansky: 'Nothing in biology makes sense except in the light of evolution'.

years of evolution, as the authors of the seminal paper announcing the first draft of the Human Genome make crystal clear.[10] This issue is all the more germane in the light of research into human DNA sequences from populations the world over. These analyses are at last beginning to bring historical order to the dispersion of modern humans throughout the globe, thereby ending one of the greatest mysteries of our common planetary past.[11] In this as in palaeontology, the South African contribution is of central importance, given the richness of the fossil remains in our country of the first *homo sapiens sapiens*.[12]

Anyone with even a passing acquaintance with the outpouring of extremely well-written popular science books on physics and biology in the past three decades will be aware of how evolutionary thinking permeates the cutting edges of new science. The question is routinely raised: how has evolution by natural selection brought about the equivalent of the computational algorithm in the human brain? Brain science itself, one of the major frontiers of current science, is permeated with the evolutionary background that has led to the as yet mysterious workings of the brain. Analogous chains of reasoning lie behind the rise of new fields such as cognitive science, artificial intelligence and complexity theory.

When one visits some of the more interesting sites on the Internet, such as www.edge.org, it is hard not to be struck by the extent to which scientists and thinkers deploy evolution as an integrating concept, and assume as a matter of course, a working knowledge of the Darwinian paradigm. Even in physics some theorists have hypothesised that evolution by natural selection may be at work in the very nature of our cosmos. The US theoretical physicist Lee Smolin has argued that the fine balance to be found among the constants of nature can only be explained by a cosmological

10 International Human Genome Sequencing Consortium. 2001. 'Initial sequencing and analysis of the human genome.' *Nature*, vol. 409, 13 February 2001.

11 The field is now an immense one. A popular summary is to be found in Cavalli-Sforza, L.L. & Cavalli-Sforza, F. 1995, *The Great Human Diasporas: The History of Diversity and Evolution*, Addison-Wesley Publishing Company: Reading, Mass. New evidence comes in almost monthly. See also Cavalli-Forza, L.L. 2000, *Genes, Peoples and Languages*, Penguin: London. For more confirmation of the African origin of not just the hominid line but specifically modern humans, see Gibbons, A. 2001, 'Modern Men Trace Ancestry to African Migrants'. *Science*, 292, 11th May.

12 For South Africa see Deacon, H.J. & Deacon, J. 1999, *Human Beginnings in South Africa. Uncovering the Secrets of the Stone Age*, David Philip: Cape Town, Chapter Six.

process of evolution by natural selection.[13] Our own universe being perhaps just one of many cosmological experiments, it is no surprise that conditions propitious to life developed in at least one. We live, perhaps, not in a universe but a *multiverse*. The natural selection of possible worlds may be the reason for the appearance of design that we see all around us, as the well-known science writer Paul Davies has argued.

In short, Darwinian evolutionary thought is part and parcel of scientific literacy in a variety of fields stretching beyond the life sciences.[14] The abstract logic of self-replication under constraints is now seen as applicable in many fields far removed from its biological origins. It has always occupied a prominent place in anthropology, despite ups and downs, especially in the United States. The idea of natural selection appeals to many economists studying competition among enterprises, while evolutionary psychology has become both an academic field and an element of popular culture. Even in sociology the trend is catching on, slowly, as the ingrained conservatism of radical sociologists begins to give way.[15]

A paradox remains: the findings of modern science are today readily available in a non-mathematical form. They are accessible to anyone who can read. This 'popular science' has now become a best-selling genre in developed countries. It makes for exciting reading, certainly more gripping and mind-expanding than many of the turgid tomes of contemporary social science. And yet here in South Africa, as elsewhere, we all know that the majority of our children find 'science' boring and incomprehensible. They avoid it where they can like the plague. Thereby they remain plunged in the ignorance that our schooling – until recently – has done little to remove.

But will Darwin last? Science, we are told *ad nauseum*, is provisional and self-correcting. As one eminent South African noted many years ago: 'Newton proved epoch-making for science,

13 Smolin, L. 1997. Weidenfeld & Nicolson: London. *The Life of the Cosmos*.

14 For an emphatic though perhaps overstated work in this regard, see Dennett, D.C. 1995, *Darwin's Dangerous Idea: Evolution and the Meanings of Life*, Allen Lane. The Penguin Press: London.

15 See van den Bergh & Fetchenhauer (2001) for a good, up-to-date review of evolutionary thinking in the social sciences, (Van den Bergh, J.C.J.M., & Fetchenhauer, D. 2001, *Voorbij het Rationele Model: Evolutionaire Verklaringen van Gedrag en Sociaal-Economische Institusies*. Den Haag-Nederland se Organisasie voor Wetenskappelijk Onderzoek).

while Darwin has become epoch-making in a far more fundamental sense. He has changed our whole human orientation of knowledge and belief, he has given a new direction to our outlook ... and has probably meant a greater difference for human thought and action than any other single thinker. But even he is not final'.[16] Smuts may well be right, indeed probably is right in some sense. Darwinism in the biological sciences has a very stable foundation, but it remains, like all scientific knowledge, provisional and subject to correction or elaboration. However, even were current evolutionary notions to be overtaken by better theory, it seems highly unlikely that they could be jettisoned entirely. Just as Newtonian physics is still the working theory outside of the very small and the very fast, so current evolutionary theory makes sense of such a vast amount of evidence that its complete supercession is improbable, though not of course impossible. Serious challenges to its fundamentals are few and far between, despite regular claims to the contrary from certain religious groupings. Within science itself, there are no alternative theories of more than a programmatic nature. The likeliest contender at the moment – complexity theory and the idea of self-organisation propounded by Stuart Kauffmann and the complexity theorists centred on the Santa Fe Institute in New Mexico – is anyway an extension rather than a replacement of Darwinian evolution, as its most fervent protagonists admit.[17]

Like it or not, and many do not like it, we are stuck with Darwinian evolution as the basis of our sciences of life. It seems foolish not to tell our children about it, and even more senseless not to take advantage of the theory's immense integrating properties in our pedagogical and public discourse.

Darwinism in South Africa: A Chequered Record

Early Rumblings: The 19th Century

Darwin and South Africa go back a long way: to 31 May 1836, to be precise, when HMS Beagle landed at Simons Town on her return

16 Smuts, J.C. 1987, *Holism and Evolution,* [first published 1928], N & S Press: Cape Town, p.185.

17 See Kauffman, S. 1995, *At Home in the Universe. The Search for the Laws of Complexity, 1987,* Penguin: London.

leg of the famous round-the-world trip that has been immortalised by Darwin's subsequent stature.[18] As was his practice, Darwin busily spent the few days ashore. He visited Cape Town and Sir John Herschel, whose conversation entertained him but whose manners he found awful. He rode on horseback to Paarl, on to Franschhoek, over the pass round to Grabouw and back down Sir Lowry's Pass to the Cape Flats. In one of those nice twists of historical irony, his very first publication was a defence of missionary activity in Tahiti, and appeared in a local journal, *The South African Christian Recorder*. There was no hint here of his later agnosticism, or of the theory that was to push Victorian Christianity into a forlorn rearguard reaction.

The dating of *Darwinism's* arrival in South Africa is not as clear as was Darwin's physical presence here. The earliest South African reference to *The Origin of Species* that the present writer has been able to trace – after an admittedly cursory search – is in the *Cape Town Mercantile Advertiser* of 28 January, 1860. Here, in an overview of international news culled from newspapers newly arrived on visiting vessels, the *Advertiser* provides a few 'paragraphs of literary intelligence'. This information includes a reference to 'Mr Murray's trade sale [which] came off on the 22nd November [1859] with a success which speaks well for the prosperity of the new literary season'. Buried in the listing of this publisher's new books is 'Mr Charles Darwin's work "On the Origin of Species" 1500 [copies sold]'. When the book itself first arrived has not been ascertained, it seems, and points to the need for a study of Darwinism in South Africa along the lines of the University of Wisconsin historian Ron L. Numbers' recent, excellent account of the American experience, *Darwinism Comes to America*.[19]

That it did arrive, and more quickly than one might have expected, is clear from a recent study by Andrew Nash (2000). The intellectual backwardness of Cape colonial society is perhaps too readily assumed. As a port of call for many ships from Europe and the USA, Cape Town was the recipient of frequent news from the metropole. Books and newspapers were eagerly awaited and

18 Darwin, C. 1989, [1839], *The Voyage of the Beagle*, Penguin: London, p.357.
19 Numbers, R.L. 1998, *Darwinism Comes to America*, Harvard University Press: Cambridge, Mass.

consumed. At any rate, by 1870 controversy around Darwin's work began to flare up in both church and educational circles. *The Descent of Man* was respectfully reviewed in one of the local serious periodicals, the *Cape Monthly Magazine,* in 1871, the same year as the book's publication. But three years later the same periodical eulogised Darwin's fierce opponent in America, Louis Agassiz, with an obituary that approvingly cited Agassiz's own attempted rebuttal of evolution by natural selection.[20] There was no unequivocal response here.

As Nash's outstanding work demonstrates, Darwin's ideas found favour not so much among the educated English-speakers but among a smaller group of Afrikaner clergy already immersed in a bitter doctrinal struggle with their conservative counterparts. With the largely forgotten but fascinating figure of D.P. Faure, founder of the Free Protestant Church (the forerunner to today's Unitarian congregation) in the vanguard, a number of clergy defended an intellectual 'Liberalism', including a critical examination of Biblical texts, that viewed religion and science as partners not antagonists. In the journal which he edited for much of its existence, *De Onderzoeker,* Faure argued that there need be no schism between scientific Darwinism and a rational, free-thinking Protestantism.[21] In this Faure was opposed not only by his conservative colleagues in the Nederduitse Gereformeerde Kerk, but also by such leading English-speaking intellectuals as Langham Dale, both Professor of Classics at the South African College and Superintendent-General of Education in the Cape, and by William Porter, Chancellor of the University of the Cape of Good Hope. As Nash comments, 'Dale and Porter's immediate task, was, in effect, to defend Galileo in such a way as to demonstrate the inadmissibility of Darwin'.[22] Thereby it might have seemed to thinkers such as Porter and Dale that both science and the received version of Christianity at the Cape could be salvaged. It was a debate that was to foreshadow similar episodes

20 *Cape Monthly Magazine,* 1870–1880, vol. 2 1871, pp.321–330, and vol. VII, 1874, pp.25–253, respectively.

21 Nash, A. 2000, *The Dialectical Tradition in South Africa.* Unpublished Doctoral Thesis, University of Cape Town, August, p.84.

22 Nash, A. 2000, *The Dialectical Tradition in South Africa,* p.86.

in our intellectual history in the following century. Its outcome too, in the defeat of the adherents of theological liberalism in the ranks of the DRC clergy, was also to be repeated.

Social Darwinism and Eugenics

Darwinism, then, first appears on South African soil as a scientific analysis that threatened first and foremost certain theological doctrines in local Christian circles, in which were to be found some of the most educated and intelligent people in colonial society. It was Darwinism *per se* that troubled the more orthodox believers among the Capetonian intelligentsia, not yet *Social* Darwinism. This ugly cousin of the theorist's work gave support to widespread folk ideologies of a *natural* basis for social and racial inequality. Again, we lack the studies that trace the detailed course of this intellectual pathogen from its home in Europe and the USA to the various regions of the then divided South Africa. It must certainly have penetrated the colonial backwaters from the 1890s onwards as science expanded in the wake of the technical demands of the Rand gold rush.[23] By the first decade of the 20th century, the idea of an evolutionary ranking of human races was clearly current in the popular mind, with references to it in the Johannesburg press.[24] Beyond popular consciousness the doctrine was at work in the minds of the country's intellectuals, *inter alia* the first major South African historian, George McCall Theal, as Dubow has shown.

Any account of Darwinism in South Africa must thus come to terms with its continued entanglement for the first half of the 20th century in the pervasive white claims for the inferiority of the black population. 'Social Darwinism' itself was based on an intellectual error that owed much to the work of the social theorist Herbert Spencer. The logical fallacy involved in the transition from Darwin to Spencer was very simple. 'Darwinian fitness' referred very precisely to the success of certain organisms to survive and *reproduce* more prolifically than their rivals in specific environments. Adherents of Spencerian notions of social fitness however failed to

23 See Dubow, S. 1995, *Illicit Union. Scientific Racism in Modern South Africa*, Witwatersrand University Press: Johannesburg, p.128.

24 See Rich, P.B. 1990, 'Race, Science, and the Legitimization of White Supremacy in South Africa, 1902–1940'. *International Journal of African Historical Studies*, 23, 4.

grasp that the crucial part of the Darwinian notion was biological reproduction. Translated into the sphere of human society, Spencer's catchword 'survival of the fittest' elided into notions of the *right* of the strongest, the wealthiest, the most intelligent.

As Social Darwinism developed in the latter part of the 19th century, it propounded a doctrine that social success was a mark of natural superiority. Its corollary – and here a dark and suspect Darwinism entered – was that the socially unsuccessful – the poor, the handicapped, the mentally less talented – were *naturally* so. Little or nothing could be done about this situation; better in fact if those at the bottom of society, or suffering from the random blows of destiny, were left to expire unaided and unmourned. This brutal creed appealed to many whose accumulation of wealth in an era of rampant *laissez-faire* capitalism led to excesses of vulgar opulence. But it missed the point – Darwin's point. Only by greater than average reproduction of their kin could any human being lay claim to higher *Darwinian* fitness. And those with the greatest claim were paradoxically the poorer classes with their notoriously large families.

Such scientific niceties were largely disregarded. Flowing out of this popular creed, but with the aid of scientists such as Francis Galton, a cousin of Darwin, arose the eugenics movement. It was dedicated to the notion that society should ensure the reproduction of its 'better' members and discourage the others from having children at all, by compulsory sterilisation if necessary. Organised eugenics was widespread in Europe and the US in the first three decades of the 20th century. It was extinguished both by the reaction to Nazi death programmes and by the incisive arguments of a number of biologists themselves, with biologists such as J.B.S. Haldane and Lancelot Hogben – professor of anatomy at the University of Cape Town in the 1930s – in the vanguard.[25]

It was inevitable that ideas propagated in the metropole should spill over and infect the political and scientific debates in South Africa itself.[26] In fact the ideas were not so much home-grown but

25 For telling anti-eugenics arguments in a lively style, see Haldane, J.B.S. 1938, *Heredity and Politics*, George Allen & Unwin: London.

26 The best account remains Dubow's book (1995), though it suffers from a rather shaky scientific grasp, (Dubow, S. 1995, *Illicit Union. Scientific Racism in Modern South Africa*, Witwatersrand University Press: Johannesburg.)

imported by the myriad scientific personnel brought to staff the new medical and science faculties of the universities. Not all adhered to the racist slant that could be given to an evolutionary account of South Africa's diverse population. But there were a number of prominent examples. The Britons H.B. Fantham, professor of Zoology and Comparative Anatomy at Wits, and J.E. Duerden, professor of Zoology at Rhodes, in the 1920s, were both strong Darwinists – and eugenicists.[27] Their *ex cathreda* notions (literally so: they were speaking in their capacity as Presidents of the South African Association for the Advancement of Science) on race rose no higher than the folk racism of the average white South African. Thus we find Duerden, the more moderate of the two in his eugenicist views, telling the assembled members of the Association in 1921 that:

> [T]he hereditary attributes of all the peoples of colour in South Africa are markedly inferior to those of the white in all that pertains to the requirements of modern civilisation, and there is every reason to expect that they will remain so in the future. For in considerations of this nature the teachings of zoology are overwhelmingly in favour of the unchangeableness of the germ plasm... In his hereditary endowments the white is far more gifted than the coloured, and must lead.[28]

But for every Duerden and Fantham, there are figures such as Raymond Dart, M.R. Drennan (professor of anatomy, UCT), S. Biesheuvel (later head of the National Institute for Personnel Research [NIPR]), I.D. Macrone (professor of psychology at Wits), and J.D. Rheinhalt-Jones of the South African Institute of Race Relations. Each of these opposed the racist conclusions of their colleagues. They insisted that the findings of modern biological and psychological science provided no warrant for the easy assumption of racial superiority.[29] And in Smuts and J.H. Hofmeyr

27 See Fantham, H.B. 1918, 'Evolution and Mankind', *South African Journal of Science*, vol. 15, and also 1927, 'Some Thoughts on Biology and Race', *South African Journal of Science*, vol. 24, 1927; Duerden, J.E. 1921, 'Social Anthropology in South Africa: Problems of Race and Nationality', *South African Journal of Science*, vol. 24, and 1925, 'Genetics and Eugenics in South Africa: Heredity and Environment', *South African Journal of Science*, vol. 22.

28 Duerden, J.E. 1921, 'Social Anthropology in South Africa', p.30.

29 See Dart, R. 1925, 'The Present Position of Anthropology in South Africa', *South African Journal of Science*, vol. 22; Drennan, M.R. 1937, 'Human Growth and Differentiation', *South African Journal of Science*, vol. 33; Biesheuvel, S. 1943, *African Intelligence*, Johannesburg. SAIRR (South African Institute of Race Relations); Macrone, I.D. 1936, 'The Problem of Race Differences', *South African Journal of Science*, vol. 33; Rheinhallt-Jones, J.D. 1926, 'The Need for a Scientific Basis for South African Native Policy', *South African Journal of Science*, vol. 23.

they found politicians whose appreciation and understanding of science were rare indeed, even in international terms. Smuts himself was well-versed in Darwin's work. He immediately appreciated the significance of Dart's announcement of the fossil *Australopithecus*. As President of the South African Association for the Advancement of Science in 1925 he gave a magisterial review of the state of natural science in South Africa in which he noted that Dart's discovery vindicated Darwin's early insight that humanity probably arose not in Asia but Africa.[30]

In the pre-WW II period, therefore, no scientific consensus existed around a pseudo-Darwinian ideology of race. Darwinists found themselves on both sides of the fence. The younger scientists coming through the ranks in the 1930s were increasingly anti-eugenicist and scornful of science harnessed to white supremacist notions. The trend continued after the war. In human palaeontology especially, South Africa continued to excel as C.K. Brain and Phillip Tobias followed in the footsteps of Raymond Dart and Robert Broom. In archaeology too a growing group of local workers expanded our knowledge of the southern African Stone and Iron Ages in a framework that necessarily took more account of the wide sweep and depth of the African past, than that employed by the historians. Archaeology and palaeontology were in this respect natural partners. Here Darwinian theory was increasingly uncontroversial and taken for granted.

But in the wider society matters stood rather differently. The pseudo-science of the eugenics movement of the 1920s and 1930s was readily co-opted by influential figures who otherwise had little or no time for its supposedly Darwinian underpinnings. One issue that united the ranks of white politicians and the majority of their constituents to an extent that is hard for us to grasp today, was that of 'race-mixture'. Inter-racial sexual relations were uniformly condemned with greater or lesser fervour. In this connection it is worth reminding oneself that more than 20 years before the notorious National Party sex and marriage laws of 1949/1950, Tielman Roos, then Minister of Justice, both passed his own

30 Smuts, J.C. 1925, 'Presidential Address to the South African Association for the Advancement of Science'. *South African Journal of Science*, vol. 22.

(milder) Immorality Act in 1927 and promised legislation to prohibit interracial marriages in the near future. (The latter never eventuated, though the issue remained a fruitful one for the 'Purified' National Party to exploit in the late 1930s as its race campaign took off.) Even in the 1940s, by which time he had long left South African shores, Professor Fantham's own declarations regarding the alleged genetic deficiencies of the offspring of mixed unions were harnessed to the cause, ironically at times by such unexpected organisations as the three Reformed white Afrikaans Churches, staunch anti-Darwinists as a matter of doctrine.[31]

Church and Evolution in the Earlier 20th Century

We are back to the clergy again. No account of the reception of Darwinism in South Africa can avoid the centrality of the three Dutch Reformed Churches in shaping state policies and popular consciousness on the issue. Again, we lack the kind of detailed studies from which valid generalisations can readily be drawn. But it seems safe to say that the history of the DRC's position with regard to the theory of evolution is a more complex one than appears at first sight. Just as the nature of Afrikaner nationalism is inexplicable without reference to the wider world (the existing world colonial system of 1900-1960, and the influence of Nazism and Fascism in the 1930s), so too must it be remembered that on this doctrinal issue the South African DRC of the time shared a common position with fellow Reformed Churches elsewhere. The struggle for the soul of the three Churches over the Bible and evolution (and more broadly, science and free-thinking) was a real one, which under different circumstances might just have taken a less dogmatic direction, as it did in later years in the Netherlands.

There is a pivotal episode here that, while not forgotten, has never received the amount of historical analysis that its importance deserves, outside of theological circles. This is of course the drawn-out battle waged in the Western Cape in the late 1920s and early 1930s over the controversial figure of Professor Johannes du Plessis, a leading light in the Stellenbosch NGK Theological Seminary in Dorp Street, Stellenbosch, and editor of a remarkable

31 Kinghorn, J. (ed.) 1986, *Die NG Kerk en Apartheid*, Macmillan, South Africa: Braamfontein, p.92.

little journal between 1923 and 1935 called *Het Zoeklicht.*[32] A former missionary and editor of *Die Kerkbode*, du Plessis assumed his post as Professor in New Testament Studies at Stellenbosch in 1916. A defender (in a biography of Andrew Murray) of the orthodox faction in the DRC during the doctrinal struggle of the mid-19th century, du Plessis had later developed a view of theology that came closer to the free-thinking liberalism of the anti-orthodox D.P. Faure 50 years before. That might not have proved fatal had he not undertaken to publish, edit, and largely write *Het Zoeklicht* for the benefit at first mainly of his clerical colleagues and students.

By the 1920s Du Plessis had become both a theological and political liberal.[33] While his politics was never openly the cause of the intense opposition to his ideas, it no doubt played a role that deserves further investigation. But he returned time and again to the issue of religion and science, in a fashion that many of his colleagues and fellow members of the NGK found impossible to ignore – or accept. Here for example is du Plessis, reproducing a speech he delivered to a student conference at the Strand in the November 1924 issue of *Het Zoeklicht*:

> Faith thus does not stand in a relationship of opposition to reason. Faith is rather just reason of a different and higher kind, or even more, reason operating in a different and higher field. ... I do not need to tell you that for this reason we must reject the distinction made by many contemporary philosophers and theologians between religious and scientific knowledge ...

And he goes on to cite approvingly from Alfred Russel Wallace's book *Darwinism* the latter's own idea of three stages of evolution culminating in human spirituality. (An ingenious resolution of the knotty problem of the place of humanity in the course of evolutionary change that caused Darwin considerable unhappiness on account of Wallace's reluctance to follow remorselessly the logic of his own – and Darwin's – theory. But that is another and earlier

32 Again the work by Andrew Nash (2000) is illuminating: see pp.105–111 of his thesis, (Nash, A. 2000, *The Dialectical Tradition in South Africa*. Unpublished Doctoral Thesis, University of Cape Town, August).

33 Decidedly with a small 'L'. As a later theologian has written, du Plessis demonstrated a benevolent paternalism together with an opposition to statutory racial discrimination: truly a 'child of his times'. (Bosch, D.J. 1986, 'Johannes du Plessis as sendingkundige'. *Theologica Evangelica*, XIX, 1.).

story.) Du Plessis's openness to new developments is nowhere better illustrated than in the February 1925 issue of *Het Zoeklicht*, in which he reacts with remarkable rapidity to Dart's discovery of *Australopithecus Africanus* in an article entitled 'The Meaning of the Taungs Skull'. Alas, du Plessis gets it both right and wrong: evolution had indeed occurred, the Biblical account of the Creation must be understood figuratively – but the Taungs find is still no proof that humanity has emerged from earlier ape-like creatures. Instead, like the Neanderthals, it could merely indicate that the human line was prone to periodic degenerations. Here du Plessis nailed his evolutionary colours to the wall, but like Broom and Wallace too, wished to keep open the possibility of a special and divine developmental path of the human species. Later DRC thinkers on evolution were to retain only the second part of du Plessis's argument: the increasing number of human-like fossils in South Africa and elsewhere after Dart's seminal discovery demonstrated the inherently degenerative nature of a sinful creature in the aftermath of the fall.

Du Plessis was to follow up his account of the Taungs skull later in the same year in the August issue of *Het Zoeklicht* with a timely article on 'The Prosecution of Scopes', once again demonstrating that the globalisation of intellectual and ideological life in South Africa is a factor too little appreciated by our historians. Du Plessis went no further than to comment that no matter what the constitutional validity of the Tennessee state law of the time prohibiting the teaching of evolution in public schools, Scopes had been correctly found guilty by the court since 'however wrong and senseless', the law had to be obeyed.

But even du Plessis's highly qualified view of evolution, along with his rejection of the literal truth of Scripture, was still too much for many of his fellow DRC members. (Not all by any means, however: the former Minister of Labour in Smuts's government, and devoted elder of the church, F.S. Malan, was to prove a staunch supporter of the Professor.)[34] By 1926 a strong conservative backlash against du Plessis had developed, and gained the support of a number of the Stellenbosch seminarians. Du Plessis was

34 See Cloete, B. 1946. *Die Lewe van Senator F.S. Malan*, Afrikaanse Pers: Johannesburg, Chapter 33.

accused of professing ideas regarding the Scriptures that could not be reconciled with Reformed doctrine. He was, in short, a heretic, his enemies claimed. In an astonishing and widely-publicised series of events between 1928 and 1931, the assault on du Plessis was rejected by the Ring of the Stellenbosch DRC, supported by a special sitting of the Western Cape Synod of the church (which had begun by condemning evolution as heresy), once again rejected in Stellenbosch and then upheld in the Synod. Du Plessis was dismissed from his post by the Church but promised his full pension should he agree to stop making public his views. He refused to do so and then took the case to the Cape Town Supreme Court, where the judge expressed his dismay at the unseemly intolerance of a Church of Peace, and nullified the DRC order. Short of entering the Stellenbosch seminary with an armed guard of police, however, there was no return for du Plessis, and he was to die in 1935 a bitterly unhappy man.

A later generation of Afrikaner theologians has exonerated du Plessis from the hostility that he had experienced at the hands of his colleagues.[35] But the immediate result of his being ostracised, according to one of these younger scholars, was nothing less than a *coup d'etat* within the NGK:

> Du Plessis represented the evangelical Reformed mainstream of the NGK and when his views were discredited as being theologically heterodox, this whole tradition which he so aptly personified lost credibility. The debacle led to much confusion and tended to stifle creative theologising for several decades. The Kuyperian trend [with roots especially among the Potchefstroom doppers but extending to within the NGK], which had never amounted to much more than a vocal and well-co-ordinated minority, took over control ... One can only wistfully speculate whether, if the evangelical Reformed position had remained dominant in the NGK and du Plessis' leadership had continued, the evolving of legal discrimination and Christian-National education may have been successfully averted in our country.[36]

While the NGK provided the backbone of Afrikaner Christianity, representing nearly 90 per cent of all enrolled DRC members, the very

35 See the issue of *Evangelica Theoligica* (UNISA), devoted to du Plessis (XXI, 1, 1988). 'Christian Higher Education in South Africa: the enterprise that miscarried'. *Theologica Evangelica*, XXI, 1.
36 Lederle, H.J. 1988. *Evangelica Theologica*, XXI, 1, pp.39–40.

much smaller Doppers wielded an intellectual, and at times political, influence quite out of proportion to their numbers. After the action against du Plessis, leading NGK personnel at the Stellenbosch seminary joined forces to multiply the Doppers' influence. The group of Gereformeerde intellectuals such as H.G. Stoker, J.Chr. Coetzee, and J.C. van Rooy for their part vigorously propagated their cause in a variety of publications, and in various influential Afrikaner Broederbond fora. Stoker and Coetzee were outspoken opponents of any doctrinal reconciliation of Scripture and modern science. They argued for a 'Christian Science' that would in all things put the word of the Bible first. (An idea which du Plessis in Stellenbosch scoffed at.) In particular, they opposed the findings of evolutionary biology with what was nothing less than an early 20th century version of 'Creation Science'. It is quite remarkable how much the issue of evolution was to turn up in the publications of this group between the 1930s and the 1950s, testifying to its symbolic centrality in the clash of ideologised theology. Just as scholars such as du Plessis could draw on one side of the great debate on evolution, so the Potchefstroom circle and sympathisers down in Stellenbosch after du Plessis's departure could refer to a growing body of work in the USA that attempted to rebut evolutionary science with science itself – for example, the 'Flood Geology' of George McCready Price and like-minded Christians.[37]

It must be said that then, as now, 'creationism' or 'creation science' tackled the pillars of evolutionary theory head-on. The following were, and in most respects still are, the main props of creationist thought; scientific acceptance of any of them would indeed be enough to bring crashing down the modern version of evolution, and many other theories besides:

- The scientific findings regarding the extreme age of the earth are faulty. Not one of the so-called proofs, ranging from biological to physical tests, is conclusive.
- Geological acceptance of Lyell's doctrine of 'uniformitarianism' is mistaken; there was indeed a world-wide flood on which the Biblical account rests, and which can account for all the phenomena that the Old Earth dogma claims to explain, like geological and fossil strata.

37 See for example Numbers, R.L. 1982, 'Creationism in 20th-Century America', *Science*, vol. 218, 5 November, p.539.

- The fossil evidence fails to show any clear evolutionary transitions between well-defined species. The record is not incomplete, as Darwin argued; it is evidence of the fixity of God-created species.
- Fossils that evolutionists see as demonstrating the evolution of human beings from a common, non-human ancestral being are either better interpreted as ape fossils or as proof of the periodic degeneration of the inherently sinful descendants of Adam and Eve. There are no 'missing links'.
- There are many anomalous findings, such as trees embedded within coal strata, that are passed over by evolutionists in a veritable conspiracy of silence, and which rebut the developmental schema proposed by geologists and biologists.
- An evolutionary process would be in conflict with the Second Law of Thermodynamics; since physicists accept its absolute validity, evolution cannot have occurred.
- The natural diversity around us, and the incredible sophistication of organic beings, transcends any possible process of probabilistic but random evolution by natural selection. The odds are much too great, and indicate the necessity for an *Intelligent Designer*.
- The origin of life itself remains a mystery. The idea of spontaneous generation from inorganic matter is, like the complex adaptations of multi-cellular beings, altogether improbable.
- Evolutionary scientists do not agree among themselves on exactly how and why evolution took place. Such intra-scientific squabbles are more evidence of the fallibility of the evolutionary schema.

At one time or another most of these objections were raised by the theologians of the DRC. Particularly influential was the prolific H.G. Stoker, Professor of Philosophy at the University College, Potchefstroom, who for more than 30 years bombarded his colleagues with both the theological and (apparent) scientific objections to evolutionary theory. His article on 'Die Ewolusie-Leer' in the 1935 collection of Calvinist essays, *Koers in die Krisis*, appears to have set the tone for much of the later pieces that lie scattered among the theological and academic literature that fervently attacked evolutionary doctrine in the name of a Christian Science.

By the late 1930s, in church and theological journals such as *Die Gereformeerde Vaandel,* (Stellenbosch), *Koers* (Potchefstroom), *Die Kerkbode* (NGK) and *Die Kerkblad* (GK), the anti-evolution position had come to stay. Both the arguments from faith – the infallible and literal truth of the Bible – and the arguments from science – the objections raised by anti-evolutionary scientists – were frequently deployed by the theologians and clergy of the Afrikaans Churches. With the Potchefstromers in the lead, the battle cry of the ultra-orthodox followers of the Dutch theologian and politician, Abraham Kuyper, *Sola Scriptura* (the Scriptures alone) replaced the humanist belief that the Bible should be subject to the same scholarly analysis as any other historical document. This 'Higher Criticism' was rejected out of hand as casting doubt on the divine inspiration of the Bible. Kuyperian 'neo-Calvinism' set the theological pace, with a minority of NGK clergy complaining bitterly about the 'verdoppering' of their church. In its stress on conformity to higher authority and its rejection of a reasoned Christian humanism, this current of theological thought fitted well with the increasingly strident nature of an exclusivist Afrikaner nationalism that pervaded the country after 1934.[38]

In their stance on evolution, the local clergy were by no means alone. It should be noted that the DRC in South Africa at this time was in accord with the doctrine of the international Ecumenical Synod of the Reformed Churches. Already in 1932, amid calls for a world-wide Reformed Ecumenical Synod emanating from the USA and the Netherlands, the Gereformeerde Kerk at its General Synod had noted that among the issues needing debate were doctrinal matters such as 'Scriptural Inspiration and Creation as opposed to Evolution-theory'.[39] Wranglings about what constituted a 'reformed' church, whether a 'Synod' of sister churches internationally could be called, and most of all the outbreak of war in 1939, prevented the convening of this Reformed Ecumenical gathering until 1949. This Synodal meeting in the Netherlands re-affirmed the historicity of the Biblical account of the creation. However, that the Reformed churches realised that this did not resolve the doubts, is indicated by

38 See for example Deist, F.E. 1991, 'Objektiewe Skrifuitleg? Kanttekeninge by Skrifuitleg in die Ned Geref Kerk 1930–1990'. *Hervormde Teologiese Studies*, 47, 2.

39 See *Handelinge van die 27ste Sinode van die Gereformeerde Gemeentes*, 1942, p.89.

their appointment of a commission into the matter shortly afterwards, consisting of leading Reformed theologians in the Netherlands. The latter argued that the strict fundamentalist interpretation was not sustainable. The South African churches submitted a 'narrow' report to this commission.[40]

In the 1930s no such international schisms were on the horizon. The Afrikaans Churches could argue that they were as sound a Calvinist movement as any world-wide. It followed from their fundamental position that if the Bible was the final and definitive authority on both matters of faith *and* of fact, then not only a Christian science but also a Christian education was a matter of clear necessity. Once again, as in the years following on from the Anglo-Boer War, the notion of Christian National Education began to resound through Afrikaner intellectual circles. The lead came most strongly from Potchefstroom.

Christian National Education and Evolution

Then, as now, the issue that most concerned identity-conscious Afrikaners when it came to education was language. Christian National Education had re-awakened in the first decade of the 20th century, in response to the Anglicising policy of Lord Milner after the defeat of the Boer Republics. (The term itself was borrowed from an ongoing struggle in the Netherlands over state funding for denominational or parochial schools that was only resolved in 1921). With state schools offering an English-language education, agencies such as the NGK had established Dutch-medium schools in the Transvaal. The struggle had been short-lived. The grant of 'responsible government' to the former Republics, followed shortly by the Act of Union in 1910, brought about a measure of dual-language medium education that placated, if not altogether satisfied, the most fiery of nationalist Afrikaners.

The upsurge of nationalist doctrine that had been building up from the late 1920s in the more ultra-Afrikaner wing of General Hertzog's National Party, however, was not satisfied with the compromise of a lax dual-medium school system that existed in

40 Du Toit, S. 1962, *Progressive Creation: An Attempt at a New Approach Towards the Problem Creation – Evolution*, Pro-Regespers Bpk: Potchefstroom, pp.42–43.

many parts of the country. What was needed was not just compulsory mother-tongue instruction for white Afrikaans-speakers, but also the promotion of the whole ethos of this ethnic section of the population. The alternative was linguistic, cultural, social, political, and, ultimately, biological extinction of the Afrikaner as a social group. With the prodding of various ideologues under the leadership of Dopper intellectuals, National Party policy after the break with Hertzog and Smuts in 1934, espoused the creation of separate school systems for Afrikaans and English language schools in which the ethos of the Afrikaner nation would be conserved. That ethos, argued the Afrikaans Churches, was linguistic and religious – and racial too, though in the climate of these colonial times the latter point did not need much persuasion.

The three Churches differed amongst themselves on precisely what policy flowed from the doctrine of Christian National Education. While the NHK and the NGK were most concerned to ensure mother-tongue instruction and a Christian outlook in schools for the children of the *volk*, the Doppers went further. They argued in favour of schools under the control of parents and the Church, which would be funded by the state, and which would expressly permit only the teaching of religious material from the doctrinal standpoint of the DRC. It was the Doppers as well who led the campaign for a Christian National Higher Education, with a long campaign for Potchefstroom University College to be allowed its own Statute that excluded the so-called conscience clause disallowing religious discrimination in staffing. In 1951 Potchefstroom gained its autonomy as a university, without the conscience clause in its Statute. In 1954 one of the leading propagandists for CNE, Professor J. Chr. Coetzee, became Rector of the University. His inaugural address, entitled 'Die Christelike Universiteit', made clear that Potchefstroom would not allow the teaching of evolution in the name of a Christian anthropology, 'which in principle rejects the idea of evolution and maintains at its core the idea of creation'.[41]

Christian National Education arrived on the national policy agenda with the accession to power of the National Party under

41 *Koers*, XXI, 5, 154, p.206.

D.F. Malan in 1948. Malan himself was a churchman, though seemingly with a sense of humour.[42] He came from the broader church stream of the NGK, and seems to have shown little enthusiasm for the more doctrinal slant of the Doppers and their fellow neo-Calvinists who were the most insistent on the need for a thoroughgoing Christianised school system. The latter were behind the production in 1948 of a little pamphlet called *Christelik-Nasionale Onderwysbeleid*.[43] It was one of the more notorious documents of the time. It contained little or nothing that had not been repeated time and again in the heady years since 1936. But since it carried the names of two Cabinet Ministers in the new National Party government, and appeared as the party came to power, it was widely taken as a declaration of the government's policy. It was, as Wally Morrow has noted, a tract infused with a kind of 'mystical nationalism'.[44] The education system was to be grounded in 'the light of God's Revelation as it is expressed in the creeds of our three Afrikaans churches ... We believe that the spirit and direction in every subject taught must correspond to the Christian and National life- and world-view ... and that in no subject may anti-Christian or unChristian or anti-national or unnational propaganda be conveyed'.

Insofar as the white schooling system was concerned, the new government proceeded less drastically than the pamphlet had demanded. The ideal cherished by the clergy of the Gereformeerde Kerk of a DRC monopoly in religious education was not pursued. It could be argued that it was not really necessary, as Nationalism had long taken over the minds of the vast majority of Afrikaner teachers. It was only under Senator Jan de Klerk – a Dopper – that legislation officially enjoining the Christian-National character of the nation's schools was passed, with the National Education Policy Act of 1967. Long before that, the Afrikaner bureaucracy, the

42 Wells, L.H., 1967, 'One Hundred Years: Robert Broom 30 November 1866–6 April 1951'. The Robert Broom Memorial Lecture. *South African Journal of Science*, vol. 63, September, tells the anecdote about Broom being awarded an honorary D.Sc. at Stellenbosch in 1946. His speech at a luncheon in his honour, with many dominees in the audience, was 'quite tactful' according to Broom himself. On sitting down, he remarked to his neighbour, D.F. Malan, 'You'll be thinking I'm almost a Nationalist', to which Malan replied, 'That's Evolution'.

43 Instituut vir Christelik-Nasionale Onderwys, 1948, *Christelik-Nasionale Onderwysbeleid*, Johannesburg.

44 Morrow, W.E., 1984, 'Philosophies of Education' in South Africa. Part 1. *South African Journal of Education*, 4, 1, p.36.

Afrikaner teacher associations and local activists had transformed the content of the curriculum for white schools. The transformation of black education proceeded of course by central fiat in a more brutal fashion with the Bantu Education Act of 1953.

Numerous authors have noted the consequences of the creeping 'Christian-nationalisation' of white schools.[45] Far more space was given to South African history from 1652 onwards, and its Afrikaner ethnocentrism was pronounced. Geography too was localised. More insidiously, as one teacher from King Edward VII High School in Johannesburg stated in 1961, 'every effort has been made in recent years to bring every aspect of school life under the direct control of the powers that be. Admissions, expulsions, promotions, failures, compulsory subjects, allocation of time to each subject, the very books one is allowed to put upon one's library shelves are now dictated from above, often in the smallest detail'.[46] As Auerbach has pointed out, the overt religious saturation of the CNE policy documents was soon overshadowed by the purely political stamp that the National Party government wished to place on the educational system.

Prior to 1948, Darwin and evolutionary theory had not been conspicuous in the relevant syllabi such as biology and botany. In the early 1930s, 'as a result of much propaganda' with Professor Fantham in the lead, a general science syllabus came into force for Transvaal high schools, including a large biology component.[47] The syllabus outline that Fantham and his colleagues had produced for the Transvaal Department of Education appears to have served as a model for the other provinces. It was intended to serve as a general science course for students up to the tenth grade, and only thereafter to allow for a biology specialisation in place of physical science.

An examination of the Cape Provincial syllabi since 1940[48] reveals that the character of the biology sections of the senior

45 See especially Auerbach, F.E. 1965, *The Power of Prejudice in South African Education*, Balkema, A.A. Cape Town.

46 Gordon, C.T. 1961, 'The Outlook for Western Education in Our Country', The Transvaal Educational News (organ of the Transvaal Teachers Association), September, p.5.

47 See Fantham, H.B. 1929, 'The Teaching of Biology in High Schools', *South African Journal of Science*, vol. XXVI, December, p.867.

48 To be found at the Department's own reference library in Bellville.

school science syllabus was not far removed from that still in force: an overview of the main bodies of knowledge considered necessary for a grounding in the subject by way of a multitude of detail. In the first two years emphasis was laid on practical work. Only in the final grades (Standards 9 and 10, or grades 11 and 12 as they are now known), and only in 1947, was the topic of evolution inserted by way of a brief study of leading biological figures; Darwin found his place among Mendel, Pasteur and others.[49]

References to Darwin began to disappear after the mid-1950s. After 1968 in the Cape, and possibly before in other provinces, the objectives of biology included: 'To develop, through contact with the subject-matter, a reverence for the Creator and an esteem for the wonders of the created universe'.[50] Prior to this no mention had been made of divine purposes in the life sciences. The addition of this objective seems to have made little change to the actual content of the syllabi, which remained heavily content-laden.

What appears to have happened was both the success of Christian National Education in the suppression of a vital part of our scientific heritage, but its failure to supplant it with a clear-cut alternative, as the older generation of DRC theologians who had urged the thoroughgoing Christianisation of the school syllabi either retired or expired. The result since the 1950s has been at school level not so much an anti-Darwinism as a non-Darwinism, reflected also in the hyperfactualism of the syllabus that the National Party had inherited from the 1940s.

Darwinism in Contemporary South Africa

The influence of this state- and church-induced amnesia is still to be perceived in the popular consciousness. Readers of the Cape Town daily *Die Burger* will be familiar with an exotic debate that has been raging for the last year or so in its letter columns and its science section. The topic has been evolution. Impressionistically at least the letters decrying Darwinism have outrun the

49 Fantam, H.B., 1929, 'The Teaching of Biology in High Schools', *South African Journal of Science*, vol. XXVI, p.874.

50 These words, or similar, appear for example in the Cape Education Department syllabi in 1973, 1978, 1990 and 1996.

pro-evolution camp (in which the science writer of the newspaper to his credit finds himself) by a wide margin.

Also symptomatic of this state of affairs is the recent book by a retired Stellenbosch architect, Pieter Pelser, entitled *Evolusie & Die Bybel*, which appeared last year and received a favourable reception in some church and other circles. Pelser's book is nothing more than a rehash of the old stock creation science arguments developed in the USA in recent decades. It is utterly lacking in originality. The book by Pelser is to the best of my knowledge the most prominent illustration of the impact of creation science doctrine in South Africa. Otherwise various flimsy creationist brochures emanate from a post box address in Durban.

Whether creation science will find an equally fertile field among South African Protestants is as yet quite uncertain. The situation here differs radically from that in the USA. Various Federal and Supreme Court judgements have ruled that the teaching of creation science under the so-called 'equal time' demand (equal that is to the time spent teaching Darwinian evolution) of its proponents, is unconstitutional. Evolutionary theory is now a respectable feature of the school syllabus in most US high schools, though local creationist pressure has resulted in its omission in some school districts in the South.

It is no longer the three sister Churches in the DRC that would be in the forefront of objections to Darwinism in our schools. Since the late-1970s the steady assault on the credibility of the DRC's old Biblical justification of apartheid has produced a new theology in the ranks of these Churches' theologians. In a remarkable about-face, the theological journals emanating from the seminaries of all three churches have moved from fundamentalist to mainstream modern Christian, with evolution no longer taboo. It is from elsewhere that opposition might come: the rapidly growing charismatic churches that are increasingly making inroads into the old congregational base of the DRC, while winning converts among non-Afrikaners too. Just how strong fundamentalist Islam is in the country is unclear. But those inclined that way are also vehement in their rejection of Darwinian evolution. The latter have adopted many of the stock arguments of their Christian counterparts, as any simple search on the Internet will demonstrate. The majority of Christian denominations, however, have learnt from the fate of the Roman Catholic Church in its attempt to dictate to science on matters of empirical fact. They know better than to take on science on science's own ground. Sooner or later the nettle will have to be grasped, perhaps by pointing out that science pursues knowledge, not wisdom. For that we need all the help we can get from believers and non-believers alike.

More to the point, perhaps, is the very folk psychology that rebels against the Darwinian message. It is interesting that in their recent research on teaching human evolution to school children on the Rand, Esterhuyse and Smith (1998) discovered that it was among high school rather than primary school children that they encountered the most resistance. This accords with the experience of university teachers, the present writer included, who find that injecting human evolution into sociology or anthropology meets with incredulity and antagonism from many, though by no means all, students. The lesson may be that the sooner we habituate our youth to the findings of modern science the better. Or it may be that adolescents of all people are those least receptive to any challenge to religious doctrine, particularly those whose own scientific training was weak or non-existent. In the long run, however, we have to teach our children not what they find most palatable, but what we think they should know.

Evolution and Schooling in South Africa

With theoretical physics at present stuck at the outer limits of current very small-scale technology[51], the most dynamic field internationally in fundamental science is without doubt molecular biology and its related fields. The centrepiece of achievement here is the 2001 publication of the first draft sequence of the whole human genome. The advances in these fields are proceeding faster than anyone anticipated, with the human genome sequencing four years ahead of the original schedule. The theore-

51 For the inability of physicists to probe as deep as they would like in order to attack the current problems surrounding quantum gravity and the ideas of string theory, see Greene, B. 2000, *The Elegant Universe. Superstrings, Hidden Dimensions, and the Quest for the Ultimate Theory,* Vintage: London.

tical and even more the practical implications are dizzying – for those who grasp what is at stake.

Under our current educational materials, that understanding is not going to come easily to our school children. We know that the universities encounter a chronic problem with the state of readiness of incoming science students. A recent study in the Western Cape found that 'approximately one in two white and Indian students counted as scientifically literate, whereas about one in four coloured, and only one in 10 African students were so classified'.[52] Students who had taken only Biology at school were at the lowest level of scientific competence. Of course there are many more reasons for this state of affairs than the content of our science syllabi. These other factors are receiving attention, though we must expect progress to be slow. But syllabus content, particularly in the higher grades, has a decisive effect on the choices made by learners, especially the brighter ones. If physics, chemistry and life science continue to be taught in the present dull fashion, then how can we expect our intake of able science undergraduates to increase?

Anyone who reads current textbooks in General Science and Biology will soon see why learners find the subjects so boring. Consider, for example, current biology texts for grades 10, 11 and 12.[53] The material is presented as a mass of overwhelming detail. One (352-page) grade 10 Biology textbook by N.P.J. van Rensburg et al. (1999), has three chapters on ecology, one on cells, three on plants, one on mammalian tissue and two on human anatomy and physiology. The underlying logic appears to be the movement among levels: from ecology to the cell, to the tissue, then onto anatomy and physiology. To a layman like the present writer, it appears quite indigestible. Why does one need to know all these facts? Nowhere is there an explanation that would make sense to the enquiring mind of the young adolescent. The grade 11 and 12

52 Laugksch, R.C. & Spargo, P.E. 1999. 'Scientific literacy of selected South African matriculants entering tertiary education: a baseline survey'. *South African Journal of Science*, 95, October.

53 See, for example, in the bibliography Van Rensburg, N.P. J, Minnie, G.E., Netshiomvani, N.M., Roux, J.S., Strydom, M.C. & van Wyk, J.P., 1999, *Study & Master Biology. Grade 10. New Syllabus;* Roedurico Trust: Somerset West, Van Rensburg, N.P. J. et al., 1996, *Study & Master Biology. Grade 11*; Roedurico Trust: Somerset West. Gibson, C. & Carrington, N. 1996, *Pass Your Matric Biology Easily*, Struik Publishers: Cape Town.

comparable textbooks are paradoxically a lot thinner. In grade 11 the order of presentation has perhaps some rationale – bacteria precede plants which precede animals – but it will be a gifted child indeed who figures out the scientific framework within which this rote-learning task is to be performed. In none of these books are the words *Darwin and evolution* to be found (I write under correction regarding 'evolution', but if it is used it is not as part of the Darwinian theory). Perhaps teachers in the better schools provide the missing key to it all, but there is no guarantee. The centrality of textbooks in forming the dominant content of teaching is a well-known and often deplored fact among educationalists. Contrast the equivalent syllabus for Australian schools, from where we are supposed to have learnt so much. Here the evolutionary background is spelt out in detail, and the underlying rationale becomes clear.[54]

If we have learnt anything from the newer trends of thought in social science in the last two decades, it is that narrative – more plainly story-telling – is the genre of choice in communication among human beings. In other words, a coherent tale with a beginning, middle and end has a structure that is readily grasped almost by nature, it would seem. Biology, history, and geography are all essentially historical disciplines, narratives of departure and arrival. One example where this narrative structure for biology has been followed with striking effect is to be found in the *Guinness Encyclopedia of the Living World*[55] written by a team of biologists mostly from King's College, University of London. This attractive volume comprises a thoroughly historical narrative of Darwinian evolution. The rationale for the material unfolds almost as of necessity. In other words, it is not as if there are no exemplars for school science writing from which we could borrow in our own curriculum reforms.

As this one example demonstrates, Darwinian evolution is both the architect and the scaffolding of our current biological knowledge. Simply on pedagogical grounds its presence in the school syllabus is indispensable. As Stephen Jay Gould has asked *apropos*

54 Information regarding this syllabus and that in the UK can be found in the *MS Encarta Encyclopedia*.

55 Guiness Publishing. 1992, *The Guiness Encyclopedia of the Living World*, Guiness Publishing: Enfield, Middlesex.

the creationist pressure against the teaching of evolution in the USA: 'Shall we deprive millions of this knowledge and once more teach biology as a set of dull and unconnected facts without the thread that weaves diverse material into a supple unity?'[56] Its absence in the South African syllabi has indeed led to the 'set of dull and unconnected facts' of which Gould complains. The consequences have been the withholding of the most powerful scientific perspectives from our youth, and thereby also very often from the public at large. What has been exceedingly harmful is the intellectual damage that the deprivation of the basic framework of biological science has caused to our children, while nourishing a wider culture of popular ignorance. Is it any wonder that biology has been the Cinderella of science schooling in South Africa, a tedious compendium of facts for children to regurgitate?[57] Biology seems to have been the choice of pupils lacking the necessary mathematical skills for physics and chemistry – a sort of scientific stand-in.[58]

As the Department of Education has accepted since 1994, our school syllabi are in drastic need not just of a purge of the worst of apartheid-inspired material, but also of modernisation. Here evolutionary thinking and findings need to find a place not only in biology but also in history, and, it might be argued, though many might think with lesser force, in physics and geography as well.

The issue is at its most controversial in the history syllabus. It is unfortunate that professional historians themselves find it difficult to begin at the beginning, and nonchalantly consign 99 per cent of our past to 'prehistory'[59]. The case is similar in the other social sciences. It is encouraging to see that reference is made to human

---

56 Gould, S.J. 1983, *Hen's Teeth and Horse's Toes*, Penguin: Harmondsworth, p.261.

57 Cf. Dekker, J.G. & van der Merwe, E.A. 1990, 'Samelewingsbiologie – 'n Voorgestelde Nuwe Benaderingswyse vir die Sekondêre Skoolkurrikulum'. In *Proceedings of National Subject Didactic Symposium*, Stellenbosch. 'There is a general feeling among [South African] biology teachers that there is a crisis in the teaching of the subject as a result of a syllabus that in great measure conveys irrelevant information'. See also Fraser, W.J. 1990, 'Kurrikulumontwikkeling en Handboekontwerp vir die Onderrig van Biologie – Wat moet wie skryf?' In *Proceedings of National Subject Didactic Symposium*, Stellenbosch, for similar sentiments.

58 Not that it helps much. Michael Kahn, (personal communication) notes that the pass rate in matric biology is lower than in physical science.

59 A point well made many years ago by the leading South African archaeologist Revil Mason in discussing the local widespread evidence of the long Iron Age in this country. See Mason, R.J. 1965, 'The Origin of South African Society'. South African Journal of Science, vol. 61, July 1965. 'The Origin of South African Society', *South African Journal of Science*, July 1965.

origins in the 2001 draft National Curriculum Statement (NCS) for the Social Sciences.[60] The NCS rightly makes the point that our region is foremost among those that have contributed to the current state of human palaeontology. But it fails to follow up this initiative with any indication of how and why this process of human evolution took place in our midst, and how it can be utilised to make sense of the *longue dureé* of world history.

School subjects all too easily degenerate into the rote learning of disconnected facts, despite the best intentions of curriculum planners. Rote-learning it must be said has had bad press. There is no substitute for it when it comes to mastering the detail of any discipline beyond the waffle of a large part of social science. At the very least however, this learning must be guided so that the learner can perceive the underlying pattern in the disparate array of facts. Among those patterns Darwinian evolution stands out for its comprehensive ordering of diverse facts and its cross-disciplinary demands. The recently released Revised National Curriculum Statement for the Natural Sciences up to grade 9 goes some way to incorporating Darwinian theory, but the test will be the design of the syllabi for the senior grades.

It can of course be argued that our younger learners yet to enter the tertiary system do not need the intellectual scaffolding provided by the major scientific theories, Darwinian evolution included. Rather, best to get some of the tough technical slog over and done with as the young brain is yet maturing. That at least seems to be the underlying assumption of the current biology syllabus in grades 10 to 12. The argument however seems dubious as soon as it is stated. It is not as if the school biology syllabus is self-evidently of practical utility in later life. Why learn a myriad details if there is no order to be ascertained and no broad conclusions to be drawn? Biology is not simple natural history, just as history is not simple antiquarianism. And what is the most efficient way of ploughing through the unavoidable rote learning of the mass of facts, if not by seeing how its various parts hang together and can be organised in mental maps of great heuristic – as well as real theoretical – value? Thus the case for Darwin in our biology syllabus is not only that it

60 *Social Sciences National Curriculum Statement 2001*, p.86.

is key substantive knowledge that should be taught, but that it is an ordering knowledge that helps the very assimilation of the material.

Evolution should, logically, be taught in *both* the natural and human sciences. In the latter sphere, Stephen Jay Gould's remark is still very much to the point: 'Evolution still floats in the limbo of our unwillingness to face the implications of Darwinism for the cosmic estate of Homo sapiens'.[61] It is precisely its implications for our view of ourselves that constitutes the crux of the problem, as Gould puts it so well here. In the natural sciences the issue could be fudged by teaching the evolutionary development of life up to but not including humanity, cowardly though it might be. No such escape route is open in the human sciences. Either we leave it out altogether, thereby leaving undefined what it means to be human, scientifically speaking, or we tell the whole story. That the inclusion of the subject is warranted on scientific and intellectual grounds is indisputable.

We are slowly beginning to accept that our education system cannot be righted all at once. The new initiative from the Department of Education, on the support for 100 special maths and science high schools in disadvantaged areas, is an encouraging sign. Differentiation of a kind is inevitable, and what better place to start with a truly modern science syllabus than in this new school grouping?

Conclusion

This paper has been written by a social scientist with no formal natural science training. I remark on the fact not so much in apology for the deficiencies of the piece, but to refer to that well-known syndrome which C.P. Snow labelled the Two Cultures, one scientific and one humanist, between which there is little fruitful traffic. South Africa is no stranger to this unfortunate cultural pathology. Phillip Tobias remarked upon it many years ago when he promoted the establishment of a science writers association in this country.[62]

61 Gould, S.J. 1996, *Dinosaurs in a Haystack. Reflection on Natural History*, Jonathan Cape: London.
62 Tobias, P.V. 1965, 'The Science Writer and Science'. *SA Journal of Science*, vol. 61, January.

Of course this is the age of specialisation; we cannot each of us range over the whole spectrum of human knowledge. And yet, speaking from the humanist side of the division, how much of compelling interest there is to learn from the other perspective! Not just interesting, but relevant too to one's own specialised concerns. There is nothing in science – apart from the mathematical apparatus – that cuts it off from the discourse in social science and artistic endeavour. *There is no good reason for the persistence of the Two Cultures.* It cannot nowadays be objected that the findings of science are beyond the ken of the average literate and inquiring reader. We now have an abundance of well-written books explaining the latest developments in the key sciences. *Popular* science, after all, is not an oxymoron.

So what prevents the development of a well-rounded popular awareness of the achievement of the human mind in both science and art? One factor may well be established religious views that cannot stomach the naturalism inherent in the scientific endeavour. It is as true now as when Max Weber wrote his famous piece on 'Science as a Vocation', that 'there are no mysterious incalculable forces that come into play, but rather that one can, in principle, master all things by calculation. This means that the world is disenchanted'.[63] Disenchantment is not the preferred state of many, perhaps the majority, of humankind. Perhaps we need to encourage the sort of intellectual schizophrenia that can entertain both scientific naturalism and the belief in numinous forces that lie behind the surface of reality with which science is concerned. Or we can move in the direction of an empirical religious consciousness that has been furthest developed in certain Asian cultures.

Be that as it may, the need is for better education, so that the experience of generations such as my own is not reproduced – to be brought up to all intents and purposes scientifically illiterate due to school syllabi in the sciences that systematically bore and frustrate the young intellect. At a time when educational renewal is a strong current in our own society, the opportunity to make scientific knowledge both accessible and exciting is there for the grasping.

63 Weber, M. 1946, 'Science as a Vocation' in Gerth, H.H. & Wright Mills, C. (eds.) *From Max Weber. Essays in Sociology*. Oxford University Press: New York, p.139.

Comment and Response to Science, Evolution and Schooling in South Africa

Wieland Gevers

There is a host of pedagogical issues tied up in the notion that schoolchildren should engage with evolution (among other demanding concepts) at the same time as they become conversant with general descriptive biology:

- Should explanatory concepts be dealt with sooner rather than later in a multi-year (Biology) curriculum? Concepts have meaning only when sufficiently large frameworks of facts and phenomena have been assimilated to *require* generalisation and systematic explanation, as a mental tool for easily assimilating further facts and phenomena iteration, is the (apparently obvious) answer, but there will still be the problem of deciding how much underpinning knowledge is required in order to introduce and consolidate conceptual thinking about it.
- Concepts tend to be more controversial than facts and phenomena, and scholars will always disagree about them. How does one arrive at a particular version of a concept that is sufficiently generally acceptable to be usable in school education? For example, natural selection can convincingly and

easily be explained as the *selective adaptive pressure* of a particular habitat on a population of *individuals* making up a particular species and displaying continuous and continual variations in their features associated with (small but significant) genetic differences. This would certainly be helpful in dealing with the more intuitive Lamarckist heresies fixed in young minds, and could take one a long way towards developing an understanding of how a species changes adaptatively, and how new species arise from it in different habitats over time.[1]

But then there is *sexual selection pressure*, arising from the workings of sexual preference (probably mostly independent of niche parameters) which can produce apparently un-adaptive change, quickly and with positive feedback, ultimately finding a resting position through the interplay of sexual selection pressure and other environmental factors.[2] Striking examples of this can be found in nature, amongst them the differences (dimorphism) between male and female phenotypes in a particular species.

A further phenomenon needed to develop a general conceptual understanding of evolution is the operation of 'arms races' between large populations of two species which are virtually at war in a single habitat, with phenotypic changes occurring in directions that mainly serve the requirements of the arms race in question, and not of the habitat *per se*.[3] The reverse of this can (perhaps) be the occurrence of static periods of evolutionary change posited in the *'punctuated equilibrium'* model of evolution, where a coelacanth can remain unchanged for much longer than a decent evolutionary process (even in an unchanging habitat) should allow.[4]

1 Williams, G.C. 1992, *Natural Selection*, Oxford University Press: New York.

2 Lande, R. 1980, 'Sexual dimorphism, sexual selection and adaptation in polygenic characters', in *Evolution*, 34, p.292–305.

3 Dawkins, R. & Krebs, J.R. 1979, 'Arms races between and within species', *Proceedings of the Royal Society of London*, B, 205, p.489–511.

4 Gould, S.J. 1982, 'The meaning of punctuated equilibrium and its role in validating a hierarchical approach to macroevolution', in Milkman, R. (ed), *Perspectives on Evolution*, Sinauer: Sunderland, Mass, p.83–104.

The fundamental issues arising from these extensions of evolutionary theory are the *direction* of change, and the principal driving forces behind them, and the *speed* of change, whether this concerns male-female differences within the species, generalised adaptive features of an entire species or the splitting off of new species from pre-existing ones.

My argument is that the world of living things, seen and experienced by teenagers, can be sufficiently clearly explained in these terms to be useful for further learning and the understanding of many extremely important phenomena of great interest and importance to them. I contend that these concepts are reasonably straightforward extensions of basic evolutionary thinking (even if over-simplified here at the risk of offending scholars working in the field and belonging to various 'camps') and should be accessible to any 15-year-old who knows the many members of the cat family, who has noted the differences between a fully grown male and female lion, and who has slept alone in the open in a predator-cat infested area.

- We next come to the 'intuitive traps' provided by the extrapolation to human affairs of evolutionary biodiversity in school curricula. Lever talks about the 'dark and suspect (social) Darwinism' derived from Spencer's mistaken association with natural selection of the idea of the survival of the fittest. Is human male-female dimorphism 'dark and suspect'? Is C.K. Brain's idea of an arms race between big-cat predators and early hominoids[5] a 'dark and suspect' contribution to the explanation of the remarkably rapid changes in recent hominoid head/human evolution? More importantly, is all evolutionary social science fundamentally suspect, not subject to verification by 'normal science'? Are these topics worth including in a Biology curriculum reaching Grade 12? If so, how? Are they sufficiently 'hard' to fix into the conceptual frameworks of millions of young people? If not, should children be kept ignorant of the *possibility* that these are amongst the

5 Brain, C.K. 1981, *The Hunters or the Hunted: An Introduction to African Cape Taphonomy*, The University of Chicago Press: Chicago.

most significant factors that have brought about the modern world in which they live?

- One must clearly also deal with the weighty and relevant questions of 'strains' or 'races' within a species. Most children have seen any number of cats (or dogs) – all one species by the definition of being able to mate with each other and produce offspring, yet obviously different. Children would also generally be aware that these differences have been achieved through breeding techniques applied by humans over the centuries. Should one explain the spontaneous generation of strains or races of humans, and what this means from a modern biological perspective? Would this be joined with some understanding of human origins and migrations, and the roles that genetics and language studies can play in explaining the modern distribution of humans in the various parts of the world?[6] The phenomenon of human language and speech-use is so central to our lives that it should be covered in Biology curricula in a way that draws on the lucid contemporary treatments of this domain.[7]
- Coming now to some of the big questions concerning the nature of humankind, one could opt in a school curriculum to follow Stephen J Gould's simple advice in his recent book *Rocks of Ages* to keep the 'magisterium' of science completely separate from that of 'morality/religion'.[8] This 'Gouldian dualism' (perhaps analagous to the 17th century mind-body dualism of Descartes) would perhaps be in order for a Marxist society but is almost certainly unsatisfactory in 2002, even for schoolgoers. Having begun properly to understand physical/biological evolution, how can one then simply throw overboard psychological/social evolution? The fact that we have triumphed in our arms race with the predator cats has much to do with our social cohesion and its associated altruism, and with our ability to learn from the past and to predict the future. It could be argued that the momentum of this and other arms races was so great that our brains, voices and hands were

6 Cavalli-Sforza, L.L. & Cavalli-Sforza, F. 1995, *The Great Human Diasporas: The History of Diversity and Evolution*, Addison-Wesley: Reading, Mass.

7 Pinker, S. 1994, *The Language Instinct*, Penguin: London.

8 Gould, S.J. 2002, *Rocks of Ages: Science and Religion in the Fullness of Life*, Vintage: London; Ballantine: New York.

prepared for the century long before our intelligence and social skills produced the tools of our technological revolution.

If our brains, and through them our aptitude and inclinations, have evolved (and who in 2002 could possibly deny that they have?) then Gould's idea could simply be interpreted as having a 'science-free' or 'magisterium' independent of the 'morality/religion' imposed on the evolved, modern, collective human mind. That is, however, tantamount to accepting, in the scientific physical/biological 'magisterium', that all people can run, dance, sing, and play chess and music equally well. Truly they can't, and Gould knew that. Humans have clearly evolved as a species to co-operate, to behave altruistically and to look into the unknown dimensions of their nature. They have evolved to enquire, to reason, to learn from each other and to take pleasure in sounds, rhythms, words and objects. But individually they vary in these attributes as much as they vary in their physical/biological features.

If one accepts this in-species diversity of otherwise generally evolved human nature, then one can begin to build a 'magisterium' to think about morality and religion, but it is a different discourse from that of Gould because it is compatible with (and seamless with the ideas) and concepts of the 'magisterium' of science. It makes possible scientific studies of altruism and kinship selection[9], of genes that affect human behaviour and the quasi-evolutionary origins of culture.[10]

Should children engage with these things or, better, should teachers be required to encourage children to engage with them? Are these things too political, too controversial, too unacceptable? If they are, are they to be silenced in the collectively educated mind of the people?

One of the lamentable difficulties of the human condition is associated with the disparity between the brilliance of scholars who can reduce the huge complexity of their fields of enquiry to a

9 Hamilton, W.D. 1996, 'The Narrow Roads of Gene Land' in *The Collected Papers of W. D. Hamilton, vol. 1 – Evolution of Social Behaviour*, W.H. Freeman: Oxford.
10 Dawkins, R. 1986, *The Blind Watchmaker*, Longman: Harlow, Essex.

simple, understandable level in books sold in up-market bookshops, and the reality of thousands of classrooms filled with restless teenagers conducting a personal arms race (not the evolutionary kind) against their teachers, the syllabus and textbook learning in general. Reading a book like *Why We Get Sick* by Nesse and Williams[11] enhances one's understanding of health and sickness as experienced by human beings, but its brilliant explanatory message will only fitfully inform textbook writing and classroom practice. Why are so many humans today short-sighted? Why have allergies become a modern curse? (Amazingly we are closer to having, but do not yet have, the true answers to these questions.)

Need the scholar-teacher divide always afflict us? Why can the Australian Academy of Science publish a market-dominating Biology textbook (*The Web of Life*)[12] and this be considered unpractical, unrealistic and even arrogant here in South Africa? With two colleagues, I was invited in 1999 to be a consultant overseeing the production of a new senior textbook series, called *Focus on Biology*, for Grades 10, 11 and 12.[13] It has 23 pages on general evolutionary theory that covers many but not all of the themes I have touched on here.

Refering now to the issues raised by Jeffrey Lever, the question about evolution in schooling today is not *whether* but *what* aspects of evolution should be taught. No one can today dispute that the basic phenomena of evolution need to be explained, illustrated and related to important real-life matters such as the natural history of HIV infection. But how do we go from there to the 'big issues' of human life on earth today that are rooted in evolution and are discussable only in its light?

11 Nesse, R.M. & Williams, G.C. 1994, *Why we get sick: The New Science of Darwinian Medicine*, Vintage Books: New York.

12 Australian Academy of Sciences 1985, *The Web of Life*, Australian Academy of Sciences: Canberra.

13 Ashwell, A., Clitheroe, F., Dilley, L., Doidge, M., Marsden, S. R., Mtombeni, G. 1999, *Focus on Biology, Grades 10, 11 and 12*, Maskew Miller Longman: Cape Town.

Religion, Science and Evolution in South Africa: The Politics and Construction of the Revised National Curriculum Statement for Schools (Grades R–9)

Linda Chisholm

This Colloquium is devoted to a discussion of the extremely significant issues of evolution and education in the context of the Africa Human Genome Initiative and the establishment of its South African branch. If one takes as a starting point that learning about the human genome and evolution are critical then three main questions present themselves. First, how does one design a curriculum in light of the constantly developing knowledge that is so much a feature of contemporary life while taking into account different styles of learning and grossly unequal resources and conditions of schooling in South Africa? Second, how does the National Curriculum Statement deal with evolution? Third, how did the public respond and what does this tell us about South Africa, the National Curriculum Statement, and the education challenges that South Africans face? This paper briefly addresses the first two, and then focuses on an analysis of the public response to the evolutionary content in the draft Revised National Curriculum Statement.

Curriculum Design in the Context of Ever-changing Knowledge

It is vital that any curriculum is designed in such a way that teachers and materials developers are able to respond to new developments in knowledge and teach it in classrooms in interesting and challenging ways. Knowledge is not static. The curriculum needs to be sufficiently broad and flexible to allow the inclusion of new knowledge as it emerges and the development of skills to access such knowledge. In the National Curriculum Statement, this flexibility of design is achieved through a statement of broad outcomes of what is to be learnt and an emphasis not only on knowledge and concepts but also on skills.

Flexibility is central to outcomes-based education, (OBE), where an active learning process is as important as the content. The National Curriculum Statement 'adopts an inclusive approach by specifying minimum requirements for all learners' (DoE 2002). Curriculum developers were explicitly asked not to overload the curriculum, thus leaving space and time in which teachers can experiment, draw on new and local resources, conduct fieldwork and engage in a range of educational activities that will supplement the minimum core to be learnt. As such, it is a guideline that is broad enough to allow for new knowledge to be incorporated by teachers and materials developers.

The Revised National Curriculum Statement is broad and flexible, but it also specifies core knowledge and skills to be learnt in each grade at ever-increasing levels of complexity. This is as a result of the view of the Report of the Review Committee on C2005[1] that learning in the gateway areas of mathematics and science, in particular, requires the systematic development of concepts – a weakness in the South African context. This is highlighted by South Africa's performance in a range of tests, in most notably the Third International Maths and Science Survey (1998) where South African students performed significantly worse overall in the maths and science literacy test. This includes even the top-

1 Report of the Review Committee on C2005. 2000, Presented to the Minister of Education, Pretoria, 31 May.

performing students (Howie and Hughes 1998). More recent tests reiterate these findings.

Social justice requires greater access on the part of the majority to knowledge and skills. As a guideline which specifies knowledge and skills from grade to grade, the curriculum aims to assist teachers in identifying the combinations of concepts and skills that need to be taught to achieve the broad outcomes. In this specification, the curriculum provides for the teaching of evolution but does so in such a manner that a balance is sought between the breadth of the outcomes and the specificity of the knowledge – skill combinations to be learnt.

Providing the minute detail is not, however, the task of the curriculum guideline. The curriculum must be fleshed out by textbooks and teachers. Teachers and texts are crucial to interpreting the curriculum and in bringing it to life in the classroom. It is necessary to leave room for teachers and texts in order to ensure that innovation, creativity and resourcefulness are built into the teaching and learning experience. This is what the curriculum aims to do: provide a sufficiently broad guideline to allow continuous incorporation of new knowledge and experiences and a sufficiently specific guide to provide the starting point for teaching and learning in schools.

Public Response to the Revised National Curriculum Statement

One of the most energetic critics of the draft Revised National Curriculum Statement for Schools (Grades R–9) was a graduate from the University of Cape Town. He worked through a Claremont-based organisation, United Christian Action (UCA), which comprised a number of religious bookshops and fraternal organisations. Its submission to the public hearing on 13 November 2001 objected to the teaching of evolution, which it considered 'nothing more than an unproven theory', and racist to boot.

United Christian Action was not alone in its criticism of the curriculum on the basis of its evolutionary content. In July 2001, when the draft Revised National Curriculum Statement was made available for public comment for a period of three months, it provoked a storm of controversy. On the one hand, departmental and ministerial fax machines were jammed with petitions, letters and appeals from a religious constituency. On the other, teacher unions, universities, non-governmental organisations, government departments and other members of the public submitted comments which supported the overall direction of the revision, but made specific recommendations for improvement.

Jeff Lever's contribution needs to be seen in the light of the comment on religion, science and evolution received from religious and non-religious quarters. Whereas Lever's paper focuses on the making of the biology curriculum in the context of anti-evolutionism and social Darwinism, the making of C2005 needs to be analysed in a wider context. Different social factors and educational philosophies have been active in it, and diverse goals, values and beliefs have supported it. This brief paper cannot unravel the full history or complexity of the process of curriculum change since 1994, but tries to cast greater light on the most recent period.

When Kader Asmal became Minister of Education in 1999, there was already significant criticism of the complexity as well as implementation of C2005, launched in 1997. Asmal rapidly initiated a review of the curriculum for Grades R–9. The Report of this Review[2] recommended that C2005 be simplified and streamlined through the production of a National Curriculum Statement. The process began in November 2001 and ended in April 2002 after the revised draft was made available for public comment. It is this revision on which this paper focuses.

This paper uses the framework provided by the work of Ball[3] and Apple[4] to illuminate the social construction of the curriculum in this historical phase. For Ball, policy has to be seen as a consequence of social struggle, negotiation and compromise; once produced, it continues to be struggled over in contexts of practice in which it is recontextualised and reinterpreted. In this view,

2 Report of the Review Committee on C2005. 2000, Presented to the Minister of Education, Pretoria, 31 May.

3 Ball, S. 1990, *Politics and Policy Making in Education: Explorations in Policy Sociology*, Routledge and 1994, *Education Reform: A Critical Post-Structural Approach*, Open University Press: London.

4 Apple, M. 2000, *Official Knowledge: Democratic Education in a Conservative Age* (2nd ed), Routledge: New York, and 2001, *Educating the 'Right' Way: Markets, Standards, God and Inequality*, RoutledgeFalmer: London and New York.

power is seen not as a struggle between dominant and subordinate forces alone, but as occurring everywhere in multiple contexts and between multiple players. For Apple[5], too, 'what actually counts [in any field of] education is a construction .... [These] constructions ... encompass multiple dynamics, multiple and partly overlapping histories and [are] in constant motion'.

Conservative religious voices were loudest in their condemnation of the draft National Curriculum Statement. But they were not ultimately the 'dominant' voices influencing the curriculum. The curriculum was shaped by a multiplicity of new, diffuse social forces, voices and educational philosophies, much less visible and loud than the evangelical Christians, but far more powerful in the new South Africa. They include environmentalist and indigenous knowledge lobbies, departmental officials, academics and teacher unionists, and the proponents of educational approaches framed in terms of perceived new national needs. Constitutional human rights and social justice imperatives and varying interpretations of the official outcomes-based education philosophy were also influential.

This paper will look first at the structure of thought and feeling of the conservative reaction to evolution in the curriculum, then at the wider response, and finally at how both were resolved in the crucible of new and emerging social forces and philosophies shaping the curriculum.

United Christian Action was not singular in its view that where the draft National Curriculum Statement failed was in its emphasis on evolution. Several hundred form letters sent from across the country saw the teaching of evolution as unChristian, 'totally unscientific', 'contrary to the Biblical account of creation' – for which 'scientific evidence' was purported to exist – and an infringement of parental rights. Blame for the teaching of evolution was attributed to the natural and social sciences learning area statements. J.C. Scheepers from Durban, for example, fulminated that the social sciences asks the teacher 'to encourage an appreciation for the significance of fossil finds in South Africa and Africa and the

5 Apple, M. 2000. *Official Knowledge.*

unfolding story of human evolution' while the natural sciences learning area statement shows 'an obvious emphasis (which ties in with the social sciences) on evolutionary processes'.[6]

Hostility to evolution in the curriculum was part of an ensemble of ideas, a structure of authoritarian populist thought, which characterised this response. Evolution was only one of the many things that bothered what appeared to be an alliance of charismatic churches, the Pestalozzi Trust (a home schooling movement led by a former University of Pretoria educationist with strong links to similar US-based movements), the African Christian Democratic Party and the New National Party. Common elements of the discourse of opposition to evolution in the curriculum were hostility to 'secular humanism', 'interfaith religion', sexuality in the curriculum and exposure of children to 'pagan' faiths and cultural practices. A common demand was the right to 'private conscience, private enterprise and family values'.

The discourse of the Christian opposition in South Africa mirrored that in the United States where 'the sheer number and range of these protests by religious conservatives exceed those by all other groups'.[7] Apple provides an analysis of the US context which is applicable to South Africa. His argument runs as follows. For many of these religious groups, morality lies in the market. As soon as the state enters, morality departs. Faith in an inerrant scriptural authority underpins the belief that people are created by God. A feeling of persecution and victimisation accompanies its reaction to change. Many feel that the secular world has declared war on Christians. To be a Christian is to be persecuted. This claiming of victimhood is a powerful rhetorical device, enhancing legitimacy at the same time as it calls for redress. It is also linked to the feeling that Christian rights are being denied as they were in state bureaucratic socialist countries. This sense of being oppressed, and yet being chosen to take the 'good fight' into the world, is fundamental to these groups' sense of themselves.

Behind it, according to Apple, lies a sense of loss – loss of faith, of imagined communities, of a vision of like-minded people with

6 Scheepers, J.C. submission, n.d.

7 Apple, M. 2001, *Educating the 'Right' Way: Markets, Standards, God and Inequality*, RoutledgeFalmer: London and New York, p.111.

shared norms and values in which the Western tradition reigned supreme. Behind it is also an intense fear of the Other and of freedom; the world feels out of control. There is a horror of mixing. On the one side is purity; on the other is danger. The danger lies in public institutions, non-believers in Christ, women, people of colour, gays, lesbians, the poor, the dispossessed. There is a strong racial subtext as well as an intense anxiety about equality.

What conservative Christians want, both in the US and SA, is public support for private schooling: voucher plans, charter schools or home schools, and the recognition of their rights and beliefs alongside others at the same time as wanting everyone else to believe what they believe. The consequence is more educational apartheid rather than less.

Not all Christians, or people of other religious persuasions, felt the same way about the curriculum and its treatment of evolution. A letter to the *Natal Witness* (28 November 2001) from the pro-C2005 Christian Network of Education noted that 'the interdisciplinary approach of C2005 offers many openings to the astute educator. Thus the natural sciences document states … that one of the unique features of this learning area is "the acknowledgement of the limitations of scientific inquiry". Not even *Christelike Nasionale Onderwys* for these Natural Science subjects made such a "religious" statement.'

The Association of Muslim Schools welcomed the values expressed in the curriculum as well as the discretion provided for schools and teachers to develop their own learning programmes. 'Equally refreshing,' it wrote in its submission to the public hearing, 'is the acknowledgement that, while science can offer solutions to many of the problems of the world, there are some problems which cannot be solved by science. For independent schools, this is particularly relevant for the acceptance of the creation standpoint as a revealed truth as opposed to the evolution theory proposed by some scientists …'.[8] For these and others, the degree of openness of the curriculum was the central issue.

8 Public hearing, Parliament, Cape Town, 13 November 2001.

The broader religious question was ultimately addressed at a political level. The Minister met with religious leaders on several occasions. In the process, the unrepresentative character of the UCA and Pestalozzi Trust became apparent, deep differences between religious denominations were revealed and confusion over school policy and curricular matters was clarified.

Evolution did not feature in the comment received from the wider public. The natural sciences learning area statement drew its strongest reaction from scientists in the university sector. Here, too, a petition was received. It focused on the depth, breadth and range of knowledge covered, as well as the need to cover the 'big ideas' of science and adequately address overcrowding, integration, human rights and the environment. There were submissions that called for more content, and submissions that called for less. A teacher union lobby sought a more adequate treatment of physics in the curriculum and an environmentalist lobby pushed for a more environmentally-conscious science curriculum. Concerns about evolution were balanced by arguments for more and better communication of science.

The subsequent revision of the curriculum occurred on the basis of principles and public comment. All working groups were required to address value conflicts in terms of the goals and principles which formed the basis of the draft Revised National Curriculum Statement in the first place: outcomes-based education, social justice and human rights, progression and integration, clarity and accessibility. It should also incorporate more awareness of a healthy environment. In relation to the Natural Sciences learning area statement, the Department recommended that the revision process should in addition reduce and broaden assessment standards, ensure coverage of major science concepts without overloading the curriculum, and address overlap and lack of alignment of social sciences with technology. The natural sciences curriculum working group was strengthened with two additional members to effect the changes. Like the other working groups, it was initially constituted through an open and public process of nomination and selection. Criteria for selection included 50/50 participation by departmental and non-departmental people, representivity, expertise and leadership in the field.

Jeff Lever's paper comes as another comment on the debate about evolution in the curriculum. There will certainly be many more papers and further debate on the curriculum in time to come. Lever's paper is important, as were those received in the process of public comment. It argues for a stronger emphasis on evolution in the curriculum. Building on the principles of outcomes-based education, the Revised National Curriculum Statement does provide space for teachers, teacher educators and materials to teach evolution through learning programmes, and to relate what they are teaching to current events, the interests of learners and evolving knowledge.

The National Curriculum can be seen as a guideline for further curriculum, teacher, materials and textbook development. It is a product of the change that has occurred in South African society where different social forces are dominant from those that shaped apartheid. Constituencies that have lost power in schooling may thus contest the teaching of evolution, but they cannot prevent it from being taught in schools. Old textbooks and approaches to teaching science and evolution may also still be in use in schools, but the conditions exist for these to change. The challenge is to ensure that they do.

Science, Evolution, Religion and Education – Creating Opportunities for Learning in South Africa's Schools

Naledi Pandor

The first eight years of democracy in South Africa have generated a great deal of debate about the form and content of education in this country. Much of the debate has focused on the massive challenge of re-organising and reshaping the administration of education, on developing a policy framework that is complementary to a democratising society, and on identifying critical elements that should make up the core curriculum in schools.

South Africa's policy makers have identified scientific knowledge as critical to developing the intellectual potential of South Africans. The primary focus in enhancing science education has been the intention to improve preparedness for post-school education in science, commerce and technology. Little or nothing has been said about science's potential to improve our understanding of ourselves and our world. Those who understand the intentions of apartheid education policy know that there was a deliberate intention to prevent access to science for black people.

Any change to this policy will therefore require an equally deliberate, focused and well-resourced set of responses. The responses will have to do far more than introduce previously excluded areas of study and exciting new 'stories'.

Lever's proposals will have to be examined in terms of such objectives and other significant imperatives that are influential in this society. He argues that schools should give attention to the study of evolution and he presents four linked propositions as support. First, he says, we need to move beyond the present dismal scientific illiteracy and ensure that we use a curriculum that prepares learners for later scientific learning. Second, science material should be made more engaging than the present array of facts that make little sense to our learners. Third, evolution as the story of humanity's origins lends itself to a narrative style and thus has the potential to be taught as a fascinating story to young people. Fourth, Darwinian evolution is both 'the architect and the scaffolding' (a clever evolutionary mixed metaphor) of biological knowledge and so should be core to the science syllabus. Put bluntly, Dr Lever holds the view that excluding evolution damages and limits learners now and that learners in the past have been damaged due to the exclusion of evolution from the curriculum under apartheid.

These propositions are not totally convincing. It is true that, under Calvinism and apartheid, evolution was not permitted to be included in the school-teaching content. It is also possibly true to assert that the absence of evolution study contributed to our poor scientific knowledge and research, but it is difficult to accept that this absence can be elevated to the status of original cause. The curriculum decisions that will eventually be made require a more convincing outline of the scientific and pedagogical rationale for such a focus.

If I were an evolutionary biologist,[1] I would make a better case in the following way: First, evolution is the only theoretical model that makes sense of pattern in biology. Sharing common ancestors

1 I am indebted to Professor Alan Morris in the Department of Human Biology at the University of Cape Town Medical School, for explaining to me the importance of evolution in these terms.

(for similarities) or having different adaptive directions (for differences) is the common theme. Second, evolution is particularly important in understanding the place of humans in modern ecology and its study should encourage us to conserve the wealth of life around us. Third, the cutting edge of biological research is in genetics and molecular biology. Genetics is evolution in action. None of us will be able to escape the implications of this research and already such terms as 'cloning' have come into common usage. Fourth, issues about race will continue to haunt us if we do not understand the meaning of racial variation and how it came about.

Upon careful scrutiny of Lever's proposals, two unusual reservations come to the fore. First, he may succeed in reviving the largely silent Calvinist lobby of yesteryear, and providing it with a large spoon for stirring up religious fervor and furore in our schools. Those of strong faith may suddenly feel they have a strong case, and that their beliefs and faith are being challenged. South African responses to religious intervention encourage a sensitive and cautious approach in the first instance.

The views of Stephen Jay Gould offer a possible response. Gould asserts that he sees no instrumental reason for the conflict between science and religion over evolution. 'Science tries to document the factual character of the natural world, and to develop theories that co-ordinate and explain these facts,' he writes. '[Religion] operates in the equally important, but utterly different, realm of human purposes, meanings and values – subjects that the factual domain of science might illuminate, but can never resolve. ... While scientists must operate with ethical principles, some specific to practice, the validity of these principles can never be inferred from the factual discoveries of science.'[2]

This is deceptively simple. Perhaps it is helpful when asserting a strongly held view in a discipline, but it is not greatly useful when dealing with the complex demand of renewing society. Many recall the enlightenment and its belief in the power of science and reason to free us from the chains of superstition and magic. Gould uses a

2 Gould, S.J. 2002, *Rocks of Ages: Science and Religion in the Fullness of Life*, Vintage: London; Ballantine: New York, p.4.

pre-enlightenment distinction between fact and value to disarm the march of enlightened thought. Facts are descriptive while norms are prescriptive. You cannot deduce normative obligations from scientific facts: there is a logical barrier between the two. That is what Gould writes and Hume wrote. This brief commentary prevents a closer look at whether contemporary philosophy would accept the distinction between scientific fact and moral obligations, as Gould puts it. However, posing the distinction this starkly does remind us of the importance of teaching moral philosophy in our schools. In the teaching of science it is necessary to refer to its potentially benign and malignant uses. Matters such as cloning, atom bombs, gas chambers and so on provide key lessons for educators and policy makers.

This brings me to my second reservation about the proposed focus on Darwinian evolution. The concept of evolution has made many positive contributions to advancing science. It has also unfortunately been used to give support to repugnant racial theories and racist movements. Lever disowns these theories and movements and refers to Social Darwinism as an 'ugly cousin' of Darwinism, or an error that supported a natural basis for social and racial inequality. He blames Herbert Spencer for Social Darwinism. Unfortunately, that is not the point. Darwin, Freud and others have all been misunderstood and have given inspiration to ugly cousins, errors and deviations.

Early writing on Darwinism points to two popular variants of Social Darwinism: evolutionism and selectionism. In the earlier, evolutionary version, the idea of mutual co-operation was more prominent than competition; the later version, which came into prominence in the early 20th century, stressed concepts of exclusion and difference, such as the 'survival of the fittest' and 'natural selection'. When Social Darwinism combined with racism and imperial ideas, it became a dangerous mixture of science and ideology.

Darwin took social ideas of his time and applied them to nature.[3] Industrial capitalism gave rise to notions of the fittest in nature,

3 Evans, R.J. 1977, 'In Search of German Social Darwinism', *Rereading German History 1800–1996*, Routledge: London, pp.119–144.

not vice versa. Thus, the issue for us now as we consider Lever's paper is this: Given our awareness that scientific ideas are shaped by the social and economic structures of their time, how do we construct a science programme that allows critical learning to stand alongside the promotion of sensitivity to the social and normative construction of scientific facts? When we look at evolution in schools in such a way, the policy makers may find they have a sound basis for placing this new understanding of society in the classroom.

ALTERNATIVE SENSE-MAKING STRATEGIES – CAN OUR SCHOOLS HANDLE THE CHALLENGE?

Bernard C. Lategan

Introduction

In his challenging overview *Science, Evolution and Schooling in South Africa*,[1] Jeffrey Lever covers a wide range of issues that call for comment and response. To mention a few: The perennial nature of the science and religion debate, the concept of 'progress' in science, the pervasiveness of science as a model for the explanation of reality, the ever present possibility of the abuse of science and of education, the inertia of school systems and of curricula, the loss of a critical consciousness and the growing importance of Africa as one of the sources of the origins of human life.

The space allocated for a response fortunately makes it easier to resist the temptation to comment on all these issues. Instead, the focus will be on a highly pragmatic question: How should we reform our school system and school curricula in the new South

1 Lever, J. 2002, *Science, Evolution and Schooling in South Africa*, Africa Human Genome Initiative: Cape Town.

Africa if we want to avoid the pitfalls of the past? Can we find a more constructive approach to the teaching of science and of religion? Surely the main thrust of Lever's argument is that the South African school system – for whatever reason – has failed to prepare students for dealing with and benefiting from the insights of contemporary science, specifically as far as the origin and evolution of life is concerned.

Rather than trying to explain why things are as they are and without underestimating the forces that will resist an alternative approach (and these are formidable!), the aim of this response is to make a few remarks regarding the basis, the framework, the elements and the strategy of a possibly more constructive and holistic mode of education for a future generation of students. Although this goes for the whole of the education system, I will follow Lever's lead by concentrating on the role of the secondary school.

Basis

Lever's analysis of the resistance to the theory of evolution provides an occasional glimpse of the subtext underpinning the intense and often acrimonious science/religion debate. The basic structure of this subtext is oppositional or exclusivist. Its logic is premised on an either/or choice: either the explanation of evolution or that of creation. The presupposition is that both options operate in the same domain and offer competing explanations of the same reality. As long as this is the case, it cannot be expected of education authorities or of school curricula to solve the problem. It also does not help to ascribe opposition to the theory of evolution to a lack of intelligence or plain backwardness. What is required is the transcription of the *either/or* subtext into an *and/and* format. Without this basic – and at the same time extremely complex – change, there is little chance of success.

What are the chances of this happening?

Narrative structure

An important key to answering the question above is provided by Lever himself when he refers to the 'consistent narrative' about 'our cosmic beginnings, our middle passage through star and galactic formation and the ultimate genesis of life on earth'. He continues: 'The story is far from complete. Parts of it may be altogether mistaken. A new twist in one part may cause ripples that reconfigure even the fundamentals on the current picture ...' It is common knowledge that accounts of creation are also presented in narrative form, like the biblical *story* of creation or the San *story* of Heisib/Xamab. The narrative form has been the subject of intense investigation by philosophers, literary theorists and historians (see for example Ricoeur 1984 & 1988; Chatman 1983; Stanzel 1982; Ankersmit 1983 and Rüsen 1997)[2]. Besides its ability to order material and to serve a communicative need, the main function of the narrative structure is its *sense-making* ability. This is especially the case when large bodies of disparate, confusing and often contradicting data are dealt with.

Darwin's theory is a classic example of such a sense-making operation in which a myriad unconnected observations, facts and experimental results is presented in the form of a narrative, that is, in the form of a process that has a beginning, a middle and an anticipated end. As Lever indicates, when the data changes or new information becomes available, the story also changes. History, that is, the recording and interpretation of the past, is essentially also a sense-making process. Countless smaller incidents and unconnected facts are structured in the form of a story that *explains* how things came about, what the trends are and what future developments can be anticipated.

Religious and cultural stories about the beginning of the world and the origins of life fulfil exactly the same function. Not all religions are classified as historical religions and not all make explicit use of storytelling, but there is no denying the constitutive and existential power of these discourses, especially

2 See Ricoeur, P. 1984, *Time and Narrative*, vol. 1, University of Chicago Press: Chicago; and 1988, *Time and Narrative*, vol. 3, University of Chicago Press: Chicago; Chatman, S. 1983, *Story and Discourse. Narrative Structure in Fiction and Film*, Cornell University Press: Ithica; Stanzel, F.K. 1982, *Theorie des Erzählens*, Vandenhoeck: Göttingen; Ankersmit, F.R. 1983, *Narrative Logic. A Semantic Analysis of the Historian's Language*, Nijhoff: The Hague; and Rüsen, J. 1997, 'Was heisst: Sinn der Geschichte? (Mit einem Ausblick auf Vernuft und Widersinn)', in Hsrg Müller, K.E. & Rüsen, J., *Historische Sinnbildung. Problemstellungen, Zeitkonzepte, Wahrnehmungshorizonte, Darstellungstrategien*, Rowohlt: Reinbeek.

in a society in which the large majority of citizens identify with one of the religious traditions. But the explanatory power of narrative structures is evident even in secular environments such as companies or organisations. When Weick exhorts managers to become historians, the aim is to make them more effective and more decisive. 'Decisive people are those who take some outcome and retrospectively construct a history that appears to have led directly to it'.[3] No wonder that sense-making has become a much-valued skill in modern organisations.

The important point for our purpose is that sense-making, often in the form of narrative structures, is an essential element in a wide variety of discourses, including the discourses of scientific explanation and of religious traditions. It is exactly in situations characterised by the overload of information, by disparate and often conflicting data, in situations of uncertainty, of rupture and radical discontinuity – be it in relation to the dominant paradigm, to the 'normalised' understanding of science or to the canonised past – that sense-making truly comes into its own.

It is this common element in both the scientific and religious discourse that provides the first building block for an alternative approach in our schools. Both provide a way to understand reality, the origins of the cosmos and of life. Each is a form of sense-making; each is a narrative with its own story line and its own integrity. The claim is not that these are discourses that are directly comparable or that they operate on the same level. The similarity lies in the sense-making function. Each of these stories should be heard and each should be told while maintaining the integrity of each discourse. Each offers its own perspective on reality and each its own way of understanding this reality. It is no solution to ignore these perspectives or to wish them away, especially in a society in which the great majority of citizens belong (nominally at least) to one of the religious traditions. But can it realistically be expected that school curricula accommodate such a diversity of perspectives?

3 Weick, K.E. 1995, *Sensemaking in Organisations*, Sage: London.

The appeal to diversity

In the present phase of South Africa's transformation, the importance of diversity as a strategic resource is increasing. The over-used reference to the 'rainbow nation' has had both good and bad press. When associated with a too shallow and too optimistic expectation of how to deal with the tough realities of the past and the present, the slogan has indeed fallen into disrepute. But it cannot be denied that substantial progress has been made in exploring the positive potential of the concept in the South African context, especially if the full spectrum of diversity is taken into account rather than only the narrow reference to racial diversity.[4] Apart from constitutional safeguards, there is a growing acceptance of language diversity, of differences in sexual orientation and of the variety of religious traditions (religious broadcasts by the national provider is a case in point). Despite its other shortcomings, the proposed curricula for 'Life Orientation' in schools are at least based on the acceptance of religious diversity and on tolerance and respect for these differences. Why should this tolerance not include the teaching of different world-views and of different sense-making discourses in our schools?

On being complementary and inclusive

The teaching of different world-views is often regarded with scepticism or rejected as being 'just not practical' The basis for this negative attitude is the prevalence of the 'either/or', or exclusivist, paradigm. In order to shift to an and/and mode, we shall have to move beyond the recognition of diversity to the recognition of the *complementary* nature of diversity, and therefore to an attitude of inclusivity.

Being complementary implies the ability to offer what is otherwise not available. Religion for one should recognise the limitations of its own sense-making abilities. It is clearly ill-equipped to offer explanations for the latest discoveries regarding the human genome. This is not its primary discourse. But it does

4 See Human, L. 1996, *Contemporary Conversations. Understanding and Managing Diversity in the Modern World*, The Goree Institute: Goree Island; and Lategan, B.C. 1999, 'Kulturelle Diversität als strategische Chance', in Hrsg J. Rüsen, Leitgeb, H. & Jegelka, N., *Zukunftsentwürfe. Ideen für eine Kultuur der Veränderung*, Campus: Frankfurt.

offer a sense-making framework to a great number of people and it comes into its own in its primary discourse of community, fundamental values and ethical guidelines. Scientific sense-making must contend with limitations of its own. By choice, it deals with specific categories of information. Although its findings enhance the understanding of reality and certainly can lead to moral implications, the moral discourse is not its primary terrain of operation.[5]

If the goal of education is a holistic one, preparing students for all aspects of the human condition, equipping them to deal successfully with the fast-changing demands of a globalised world and sustaining the hard-won democracy in South Africa, a complementary approach becomes a necessity, not a luxury. All that the Natural Sciences have to offer should be unrestrictedly part of it, as should be the Humanities. Religion is but one part of it. History and historiography as sense-making operations *par excellence*, the explanatory contribution of the social sciences and the rich legacy of the humanistic tradition, including the African understanding of society and humanness – all have crucial contributions to make. It is for this reason that the present allocation of time and attention to Life Orientation in the school curricula is totally inadequate for the purpose. In its present form it has no realistic chance of contributing significantly to 'rebuilding the moral fibre of South African society'. It is therefore highly irresponsible if not hypocritical to bemoan the decline of moral standards without making adequate provision in school curricula to address these very issues in a substantial manner.

It would be an illusion to think complementarity means a seamless fit. The radically differing nature of these discourses has already been pointed out. But it is also a mistaken notion that religion and science have nothing to do with each other and that they can be kept neatly apart. Both are trying to make sense of the same 'reality'. To be complementary in this situation means to respect the integrity of the different discourses and the different goals they pursue. The need to hear both is all the more important,

5 See Ridley, M. 1996, *The Origins of Virtue*, Penguin: London, for an interesting attempt at explaining human social nature from the perspective of evolutionary biology.

as both science and religion have a dark side. Lever illustrates for us how an ideology that is religiously motivated can seriously damage a whole education system for generations. It is important to realise that this tendency is not limited to one brand of religiosity. Lever shows how effectively a certain type of Calvinism was used by the Reformed Churches in South Africa to give moral sanction to a political ideology and to keep school curricula free from teaching any form of evolution. But the same kind of resistance can be expected from all types of religious fundamentalism, be they in a Christian, Muslim, Jewish or even an African religious mould.[6] Science, for its part, is equally open to abuse, as Lever illustrates in his account of the dubious history of eugenics.

A complementary approach to the teaching of science and religion therefore also implies a non-absolutistic and a non-proselytising attitude – which in essence means teaching with appropriate humility. What sense-making discourses offer is a way of interpreting the available data or information as these stand at the moment. New discoveries, new insights and new circumstances would produce new insights that may require adaptation and revision.

The return of a critical consciousness

A complementary, inclusive, non-absolutistic approach could provide the conditions favourable for a return to a critical consciousness – that quality that is so desired as an outcome of our schooling system and that is so dearly sought after by tertiary institutions, and is so often not achieved. This quality is not foreign to our own tradition, as Lever reminds us by referring to the Du Plessis saga.[7] 'Critical' in this context does not mean destructive, but informed, balanced, independent and courageous.

Is an alternative approach viable and possible?

To return to the initial question: What are the realistic chances that an alternative approach to the teaching of science and religion will be implemented in our schools?

6 Castells, M. 1997, *The Power of Identity*, Blackwell: Oxford, pp.6–67.
7 Lategan, B.C. 2001, 'Preparing and keeping the mind-set intact', in *Scriptura*, vol. 76.

Most of the basic prerequisites are in place. The recognition of a plurality of world-views and respect for different religions are already embedded in the proposed curriculum for Life Orientation and the teaching of religion. The acceptance of alternative understandings or explanations of reality, based on scientific observation and information, lies in the continuation of this recognition. The complementary nature of these alternative sense-making strategies is accepted even by religious bodies, as is evident in a recent statement[8] of the leadership of the DRC in the Western Cape in this regard – provided the integrity of each discourse is retained. This approach is further strengthened by other forms of plurality and diversity in South African society, apart from constitutional safeguards. The necessary critical consciousness is part of our own past, even if it is in need of re-awakening.

The critical question is whether the relevant role players will be willing to implement and sustain an alternative approach – and if they are, whether the teachers and material will be available and prepared to deal with this much more demanding and challenging responsibility. If we owe it to a new generation of students to prepare them more adequately for a new set of challenges, it becomes – in the final analysis – the responsibility of us all.

8 DRC Moderature 2002, 'Kerk en wetenskap', in *Die Burger*, 13 February.

Evolution, Creationism, Indigenous Knowledge

2

The Evolution/Creationism Debate: Insights and Implications from the Indigenous Knowledge Systems Perspective

Catherine Odora Hoppers

Introduction: The Gould and Lever Contributions

The key materials availed for the preparation of this paper are Stephen Jay Gould's book, *Rocks of Ages: Science and Religion in the Fullness of Life*[1]; and a paper by Jeffrey Lever entitled *Science, Evolution and Schooling in South Africa.*[2]

The position taken in this paper is that both materials offer invaluable analysis on the nature of the tension between the evolutionist goats and the creationist sheep with each author taking fairly lucid and convinced positions as to where to go from here: for Gould – Non-Overlapping Magsteria (NOMA), and for Lever, evolutionism 'for ever'. The two materials also present the reader with fascinating arguments to support both positions, with

1 Gould, S.J. 1999, *Rocks of Ages: Science and Religion in the Fullness of Life*, Vintage: London; Ballantine: New York.

2 Lever, J. 2002, *Science, Evolution, and Education in South Africa*, Africa Human Genome Initiative: Cape Town.

Gould taking a firm referee position and holding the arbitrator's hammer and saying 'enough!' to the seemingly endless battle for territory by the goats and sheep alike.

Gould's intellect is at a fine level indeed as he writes:

- On diversity: ... *we must live in the fullness of a complete life in [the] many mansions of a neighbourhood that would delight any modern advocate of diversity.*[3]
- On the nature of the realms: ... *science is the age of rocks, religion – the rock of ages. Science studies how the heavens go, religion focuses on how to go to heaven.*[4]
- On self-restraint by both realms: ... *if religion can no longer dictate the nature of factual conclusions residing properly within the magisterium of science, then scientists cannot claim higher insight into moral truths from any superior knowledge of the world's empirical constitution.*[5]
- On the link between the two along a continuum: ... *God had indeed created nature at some inception beyond the grasp of science; but he also established invariant laws to run the universe without interference thereafter* .[6]

He advocates a principled domain separation – a respectful non-interference by stating that science documents the factual character of the natural world and seeks to explain these facts (the empirical realm), while religion operates in the realm of human purposes, meanings and values. Gould's NOMA recognises that science cannot answer all questions and cannot dictate social policy,[7] and that religion has fostered the most unspeakable of horrors, including inquisitions and liquidations, but has also brought out incredible human goodness.

Lever examines the stunting impact that civil religion of ethnic nationalism (Christian National Education) had on freedom generally, and on evolutionism in particular, and reiterates the

3 Gould, S.J. 1999, *Rocks of Ages: Science and Religion in the Fullness of Life*, Vintage: London; Ballantine: New York, p.4.
4 Gould, S.J. *Rocks of Ages*, p.6.
5 Gould, S.J. *Rocks of Ages*, p.9–10.
6 Gould, S.J. *Rocks of Ages*, p.21.
7 Gould, S.J. *Rocks of Ages*, p.169.

case for science. Over the last 150 years, Lever argues, modern science has assembled a remarkably coherent picture of our universe and humanity's place within it. He celebrates the advance, together with technology, of science and the scientific revolution of the last 200 years:

> Thus science and technology advance in tandem. Our technology rests on our understanding of the scientific fundamentals that are derived from the overarching theories brought into fruition in the 20th century. The technology in turn deepens our theoretical progress in a continual feedback loop.[8]

Apart from ground-breaking achievement and off-shoots such as brain science, the human genome project – the 20th century scientific endeavour – according to Lever, has abolished once and for all the mechanistic determinism that Newtonian physics seemed to instill in the universe. Materialism now rests on shaky foundations. We live today not in a universe but in a multiverse. To be further and positively noted is the wide permeation of Darwinian evolutionary thought in many fields far removed from its biological origins (recall anthropology), especially among economists and sociologists.

From the above, Lever acknowledges the use to which Darwinism has been put in the context of racism, but none the less present an evangelical case for evolutionism arguing that 'Darwin's account remains intact, stubbornly uncontroverted',[9] and wonders how such exciting issues could not be called on to liven up the science curriculum which children find 'boring and incomprehensible'.[10]

Views from a radical witness and wounded healer[11]

Core to this paper is the basic acknowledgement of important advances in the fields of science and technology and the equally important place that evolutionism has in the scheme of things.

8 Lever, J. 2002, *Science, Evolution and Schooling in South Africa,* Africa Human Genome Initiative: Cape Town, p.3.

9 Lever, J. 2002, *Science, Evolution and Schooling in South Africa,* p.4.

10 Lever, J. 2002, *Science, Evolution and Schooling in South Africa,* p.8.

11 See Richards, H. 1995, *Education for Constructive Development.* The Nehru Lectures given at the University of Baroda: Gujarat State India. August/September 1995.

I personally also agree that the way science is taught in general, but in non-western societies in particular, leaves much to be desired. Likewise, Gould's NOMA is acknowledged as an elegant conflict resolution instrument, as well as there being far too little traffic between the scientific and humanistic (to which one can also add the spiritual) cultures.

Is the introduction of evolutionism/Darwinism the real *question*, or for that matter, the *answer* to the issues that concern Lever? Given the painful and deeply segregated history of South Africa, where is the place for diversity, and the link between epistemology and democracy in all of this? Equally disturbing are silences in response to the distasteful if not destructive role that science has played in legitimating the wanton destruction of nature, sanctioning and reifying our disconnectedness with it, and the consequences that have flowed from this. When this latter concern is taken into account, one can ask whether the question before us should still stand as fossilised as that of the battle between religion and science.

As a radical witness and a wounded healer, I find that the celebrations to which Lever invites the reader, and the dispute to which Gould's attention is focused, both look too benignly at the two fields of religion and science. Both authors, it can be said, are preoccupied with power and management issues between the two, i.e. why the 'bloodshed must stop', and for Lever, why a new day must surely dawn on evolutionary theory so it can take its rightful place in contemporary history. Both are either naïvely or dangerously oblivious to the manner in which these two selectively 'couple up' when called upon by terrestrial forces such as imperialism, colonialism, modernisations and apartheid.

When this factor is introduced, the picture begins to closely parallel what Woddis,[12] referring to the continuing obscene structure of the present world order, referred to as the *neo-colonial coalitions* – the 'bedfellows' – that make possible the joint exploitation of the former colonies. These joint efforts, begun way back on the day Queen Isabel and King Ferdinand granted

12 Woddis, J. 1967, *Introduction to Neo-Colonialism*, International Publishers: New York.

Christopher Columbus the privileges of 'discovery' and 'conquest'[13] were sustained throughout the colonial and post-colonial period. In South Africa, a vicious form of it was to be presented by the indigenous population.

The tripartite consortium between science, religion and imperialism, which Woddis called 'collective or consortium colonialism', the 'unholy condominium', has to be looked at closely and with fresh lenses in order for the impact it has had on the 'other' to be grasped, especially the unanswered charge of 'attempted murder' on non-western peoples' knowledge and heritages in the ways discussed below.

Developmentalism: The imposition of the western monoculture over others

Working in tandem, this tripartite consortium betrayed the principles of democracy and humanism. It pre-empted the possibility for the study of inter-subjective grounds of human action, and for us to take cognisance of the polysemy of history, its many meanings, its possibilities and its ambiguities.[14] It mutually reinforced the entrenchment of development paradigms that were not only transplanted badly, incorrectly, and unjustly', but which transmogrified billions of the world's majority into an inverted mirror of western identity, a mirror that belittled them and sent them 'to the back of the queue', a mirror that defined their new identity. For a heterogeneous and diverse majority, the mirror jaundiced their identity simply in terms of a homogenising minority.[15]

Developmentalism, or the theory of linear progress which correlates with specific epochs of western hegemony, found its particular enhancement through the comparative method whose underpinnings succeeded in theoretically incorporating non-western societies into its paradigm. Essentially a discourse of power and subjugation, it is constituted in a context of patriarchal global relations as a *recipe for social change.*

13 Shiva, V. 1997, *Biopiracy: The Plunder of Nature and Knowledge,* South End Press: Boston. p.1

14 Latouche, S. 1991, *In the Wake of the Affluent Society: An Exploration of Post Development,* Zed Books: London.

15 Esteva, G. 1992, Development. In Sachs. W. (ed.) *The Development Dictionary: A Guide to Knowledge as Power,* Zed Books: London. pp.6–7.

...The central thesis of developmentalism is that social change occurs according to a pre-established pattern, the logic and direction of which are known [from European experience]. Privileged knowledge of the direction of change is claimed by those who declare themselves furthest advanced along its course. Developmentalism is the truth from the point of view of the centre of power; it is the theorisation (or rather ideologisation) of its own path of development, and the comparative method elaborates this perspective.[16]

From the point of view of the centre, global space is transformed into a time sequence, with Europeans as the only contemporaries, the sole inhabitants of modernity; a perspective which served successfully as a manual for imperial management of societies 'at different evolutionary stages'. Africa, needed as unpaid subsidy for western development, was put at the bottom of this schema. Within this framework, Europe (now read the 'West') defined the world and gave names to phenomena in the genesis of the new world society brought forth in the wake of European expansion and conquest, industrial revolution and the advancement of the world market. The naming process itself was an extension of the process of conquest-making; becoming 'modern' meant becoming western.[17]

Following on fast tow, the term 'underdevelopment' soon took on an unsuspected colonising virulence of its own, making development connote at least one thing: to escape from the undignified form called 'underdevelopment'. The point here is that, in order for someone to conceive of the possibility of escaping at all from this 'condition', it is necessary first to feel that one has fallen into that condition. And for those who make up two thirds of the world's population today, the pressure to 'feel' this condition is exerted in a regular manner in the very process of executing development.[18] In order for the so-called underdeveloped to think of and comprehend 'development' as stipulated, expected or

16 Pieterse, N.J. 1991, 'Dilemmas of Development Discourse: the Crisis of Developmentalism and the Comparative Method'. In *Development and Change*. vol. 22, Sage Publications: London, vol. 5–29, p.2.

17 Pieterse, N.J. 1991, 'Dilemmas of Development Discourse'.

18 Sachs, W. 1992, Introduction. In Sachs, W. (ed.) 1992, *The Development Dictionary. A Guide to Knowledge as Power*, Zed Books: London. pp.1–5.

demanded by the West, they are assisted, by means of various statistical and other forms of written material emanating from the West, to form and internalise the perception of themselves as underdeveloped, along with the whole burden of connotations that this carries.

As the mental space in which people dream is occupied by western imagery, the innumerable varieties of 'being human' are eliminated. However, as the 'other' has vanished with the arrival of development, the spreading monoculture continues to erode viable alternatives to the reductionist, exploitative paradigm of society reified by science, and to cripple humankind's capacity to meet an increasingly changing future with creative responses.[19]

Scientism

What we need to do is to distinguish the performative aspects of science, i.e. the creations, the technologies and discoveries associated with it, from *scientism*, which is the ideological use of 'science' defined Eurocentrically as an activity which sanctions all thought and behaviour, and posits science as sacred, the highest standard of morality. It is this ideological use of science that has objectification as a key cognitive modality; and it is this objectification that designates everything other than the 'self' as object.[20]

Science turns the 'self' into an ultimate manipulator, a spectator, a recorder of events, an act of alienation. Over time, the gaze of science over the so-called 'non scientific' became a gaze of power, and of surveillance. Visvanathan states that the way science is constituted prevents the entry of pain and compassion, leaving the 'I' of science an impoverished self without a backstage. Science, he argues, is not only political, but goes beyond politics to create its own micro-physics of power, with its own capillaries, by pre-empting and terminally judging the way a person thinks.[21]

19 Sachs, W. 1992, *The Develolpment Dictionary*.

20 Visvanathan, S. 1997, *A Carnival for Science: Essays on Science, Technology and Development*, Oxford University Press: Calcutta.

21 Visvanathan, S. 2002, 'Between Pilgrimage and Citizenship: The possibilities for Self-restraint in Science'. In Odora Hoppers, C.A. (ed.) 2002, *Towards a Philosophy of Articulation: Indigenous Knowledge and the Integration of Knowledge Systems*, New Africa Education: Cape Town, (forthcoming).

The Baconian edict had itself already shifted the goals of science from wisdom to manipulation, and brought together the two strands of the *mechanistic* conception of reality, and the *male obsession* with domination and control. It was Bacon who advocated an empirical approach to science that was passionately vicious.

'Nature,' Bacon said, 'had to be hounded in her wanderings, bound in service, and made a slave. She [note the pronoun] was to be put in constraint [and the work of the scientist was to] torture nature's secrets from her'.[22] Similar words have been commonly used during the period of colonial penetration and defilement to express what was to be done to Africa (whose virginity was to be wrested). It is also this confluence of attitudes that makes anything 'African' to be posited as an object of research, to be exploited without recompense in the name of science.

The scientific method, combined with the colonial enterprise, thus stands implicated in the worst excesses of imperialism. The term 'research' itself inspires in local communities a silence, and conjures up bad memories that still offend the deepest sense of our humanity. It galls non-western societies that western researchers, intellectuals and scientists trained in that tradition can claim to know *all* there is to know about the latter, on the basis of brief and superficial encounters with those societies. It often appalls indigenous societies that western science can desire, extract and claim ownership of people's way of knowing, and simultaneously reject the people who created those ideas and deny them the opportunities to be the creators of their own culture and notions.[23]

Researchers in the area of Indigenous Knowledge Systems should, in particular, decry the manner in which the social sciences have joined the natural sciences in assuming that the subject (researcher) can be separated from the object (person

22 Merchant, C. cited in Capra F. 1988, *Uncommon Wisdom. Conversations with Remarkable People,* Flamingo: London. pp.226.

23 Smith, L.T. 1999, *Decolonising Methodologies. Research and Indigenous Peoples,* Zed Books: London. See also Odora Hoppers, C.A. 2001. *Decolonizing the Curriculum, IKS, and Globalisation.* Document prepared for the Gauteng Institute for Curriculum Development and the CEPD. April 2001.

being researched) by a methodological screen.[24] It is more than apparent the way academic social and natural sciences research continues to exploit the poor and the marginalised, especially blacks and women, in the name of science. Weskott cites the case of black people in the United States who were measured, analysed, processed, dissected and reduced to manipulable data that advanced the career interests of the investigators, but did little to improve the plight of the investigated.[25]

Most of the time, women have been an attractive subject to exploit so long as the purpose of social knowledge is simply to obtain more information. In social science's unrestrained pursuit of information, any new object of study that can generate mounds of data becomes 'very interesting', and is of interest so long as it is a prolific source. When the data is no longer new, the object loses its primacy and value.[26]

The imposition of the obligation of acceptance on non-western societies

Latouche has argued that as the vitality of a culture resides in the capacity of that culture to *give* (both symbolically and materially), receipt of a gift (whether this be willing, inadvertent, or forced) is *prima facie* evidence of *its valorisation by the recipient*. The gift, and the capacity to give and to have the gift received, signifies the existence and potency of the donor as an active agent in the world.[27]

As it continues to stand, the West's primary domination of the world lies in its monopolisation of the very *terms* by which value is conceived, and its domination of the basic institutions that codify social life.[28] The *deculturation* of the dominated societies is shown by the fact that, increasingly, they exclusively voice their predicaments and aspirations in terms of the categories sanctioned by the invading culture. Western culture has imposed

24 See Weskott, M. 1987, 'Feminist Criticism of the Social Sciences'. In *Harvard Educational Review*. vol. 49. no.4.

25 Weskott, M. 1987, 'Feminist Criticism of the Social Sciences'.

26 Weskott, M. 1987, 'Feminist Criticism of the Social Sciences'.

27 Latouche, S. 1991, *In the Wake of the Affluent Society: An Exploration of Post Development*, Zed Books: London, p.10.

28 Latouche, S. 1991, *In the Wake of the Affluent Society*, p.10

the *obligation of acceptance* on the invaded cultures. This entails, at the limit, the asphyxiation of the recipient culture, and the loss of vitality and coherence of the indigenous cultural forms. The Third World societies are, under these conditions, made to feel that there is little, or nothing, that they have ever given to others.

This most basic 'unequal exchange' in the Centre-Periphery relations, is the fundamental logic behind their description as poor and backward, irrational, etc. It is the devaluation of non-Occidental societies in this way that is both the price *and* the precondition of their entry onto the path of 'economic development'. In short, underdevelopment is a process of genuine deculturation without the material (and symbolic) benefits of the complementary or reciprocal acculturation. It signifies the *cruel inscription* of entire societies to what is, in effect, only *nominal salvation.*

Indigenous Knowledge System (IKS): The subjugated knowledge

Indigenous Knowledge Systems represent the systems of knowledge in philosophy, medicine, science, technology, religion, education, metallurgy, etc. of non-western societies. These knowledges, summarily rendered 'irrelevant to their use' (in Foucauldian terms) by the unholy condominium, are underpinned by a way of life, a way of seeing and a way of relating that brings to the picture a different set of propositions:

- As opposed to the Baconian edict, the IKS stresses instead the essential interrelatedness and interdependence of all phenomena – biological, physical, psychological, social and cultural. Indigenous cosmology centres on the co-evolution of the spiritual, natural and human worlds. Thus many indigenous peoples in Africa still practice the ritual of burying their babies' umbilical cords and immediately planting trees on the spot in order to establish a relationship with the plant life. Family histories make reference to some animal totem to be conserved. The IKS holds that there are sacred places in the wild that have to be avoided but also conserved. These are places where people are not permitted to fell trees, hunt wild life or collect wild fruits for commercial purposes. Natural phenomena such as rivers and mountains are not seen purely as objects whose worth is calculated in currency. They play a significant role in the psyche and constitution of African people.

- As opposed to the linearity and reductionist impulses of the contemporary development approaches, paradigmatic imperatives in an IKS-conscious development emphasises a people-centered approach and actively tries to shift the paradigm and discourse of development from one that is pre-occupied with what the people *do not have,*[29] one that is trapped in a negative dependency orientation that it generates, to one in which people are the *subject,* and which motivates society to become constructively engaged in moving forward. The endogenous development to emerge from this process begins at the point when people start to pride themselves as *worthy human beings inferior to none*; where such pride is lost, development begins at the point at which this pride is *restored,* and *history recovered.*
- The IKS brings to the fore the ethical underpinning that academic practice should not condone or encourage the *further trapping of the poor in the cold, condescending gaze of the rich.*
- The IKS requires that the Unesco Agenda for Science in the 21st Century (Budapest 1999) is put to active implementation. The Science Agenda puts forward the important caveat that there is a need for a vigorous informed, constructive inter-cultural and democratic debate on the production and use of scientific knowledge. It urges the scientific community to open itself to a permanent dialogue with society, especially a dialogue with other forms of knowledge.[30] It affirms that modern science does not constitute the only form of knowledge, and closer links need to be established between this and other forms, systems and approaches to knowledge for their mutual enrichment and benefit, and in order that better ways are found to link modern science to the broader heritage of humankind.[31]

29 Rahman, A.M.D. 1993, *People's Self Development. Perspectives on Participatory Action Research,* Zed Books: London. p.216.

30 *Science for the Twenty First Century. A New Commitment. The Declaration on Science and the Use of Scientific Knowledge.* Paris. UNESCO 2000. pp.21

31 *Science for the Twenty First Century,* UNESCO, pp.27.

- The IKS draws attention to cognitive justice, human rights and inclusivity in national innovation systems; it calls for a fuller development of cognitive praxis between various domains, and sheds light on innovation undertaken by communities now forcibly stigmatised and relegated to the 'informal' sector.
- IKS also brings together various threads in international customary law. The Universal Declaration of Human Rights (UDHR), for instance recognises the inherent dignity, equality and fundamental rights of all people in all nations. This single, most authoritative source of human rights ensures the right of people to freely participate in the cultural life of the community, to enjoy the arts and to share in the scientific advancement. These rights are further emphasised by Article 15 of the International Covenant on Economic, Social and Cultural Rights (ICESCR 1966), Article 19 of the International Covenant on Civil and Political Rights (ICCPR 1966) and the Vienna Declaration and Program of Action (VDPA 1993).[32]

Conclusion

In stating once again that the nature of the battle itself needs to be redefined, the following five propositions can be outlined:

- Perhaps we need to look beyond the tension between evolutionism and creationism to the reconstruction of a future that must result in a new whole that is greater than the sum of its parts. We need a future that focuses on the transformation of world – views and ethics of humankind, cognitive praxis and the creation of a third interstitial space[33] in which knowledge can be reframed and social organisation of trust can be negotiated.
- Perhaps we need the African perspective now: the *reasoning embrace*, the sympathetic reason.[34] The core African philosophy of Ubuntu would then be practised not as an effort that is applied sporadically, but as a constant *turned-*

32 See WIPO/UN Commission for Human Rights Conference Roundtable Report 1999.

33 Gough, N. 2000, 'Globalisation and Curriculum Inquiry: Locating, Representing and performing Transnational Imaginary'. In Stromquist, N. & Monkman, K. 2000, *Globalisation and Education: Integration and contestation Across Cultures*, Rowman & Littlefield: Maryland. pp.77–98.

34 Shutte, A. 1993, *Philosophy for Africa*, University of Cape Town Press: Cape Town. p.16.

towardness, a *human-tropic* dynamic of relational living (just like photo-tropism in which plants benefit from sunlight).[35]

- Perhaps, instead of focusing on the battle between the domains of religion and science, we should look at patterns behind the patterns and processes beneath the processes; introduce history and culture and aim at a socio-economic and spiritual upliftment within a project of enlarging the metacognition of all humanity.
- Perhaps clues to be pieced together in the re-animation and humanisation of science, especially the reconstruction of its relationship with those numerous suppressed 'others' – women, nature, the IKS and the non-West – will take us one step higher than the present transfixion with its combat with religion. Should the question then not be how to help science recover its sense of dwelling, its sense of caring, crying, laughter and joy? Should we not be concerned at the methodological pomposity of science; break its brittle authoritarianism and have it abandon the dowdiness of the vivisectional mandate and the triage inherent in its practices; recover its conscience, its role as an agent of plurality, of heretical dissent, and its union with nature? When *anima* is restored to nature, won't that respiritualise science, deepen it and lead to it the satyagrahic experience that brings together the cognitive, the political and the spiritual?[36]

It has been said that every age has a prototypical violence. The violence of our age is based and feeds upon our indifference to human suffering, to our ecology and to the origins and sources of the life from which we derive our existence. What we need is therefore not a renewed battle to the death between the two 'sects' – religion and science – but an *ethical stance*, which is anchored in the possibility of an authentic dialogue between cultures that could lead to an equally authentic co-existence of different cultures. This may imply that certain cultures renounce their barbary in order to

35 Malan, J. 1997, *Conflict Resolution Wisdom From Africa*, Accord: Durban.

36 Visvanathan, S. 2002, 'Between Pilgrimage and Citizenship: The possibilities for Self restraint in Science'. In Odora Hoppers, C.A. (ed.) 2002, *Towards a Philosophy of Articulation: Indigenous Knowledge and the Integration of Knowledge Systems*, New Africa Education: Cape Town, (forthcoming).

have the other renounce its own. In the meantime, the hope is to propagate willingness to tolerate contradiction, and to act generously in situations of *unresolved antagonism*.

From the perspective of those that have been wounded by both sects, a central concern is that, over the past 200 years, the West has exercised a monopoly over the very definition of the problems of existence of all societies in the world. The crucial challenge for those in the West wishing to afford genuine respect to those 'other' than themselves would be to quit the paranoia, relinquish this monopoly and listen more openly to the discordant messages of silenced populations. The search for reciprocity – of a *'space of fraternal co-existence'* – is itself an ethical choice which carries some serious implications. It implies affirming the richness of the other, even in their material poverty. It implies affirming that this is not a matter of quantity but quality of life, and that all helping is reciprocal, just as learning must be reciprocal. Respect for the other implies acceptance of dissension, of loss and of death, which is translated into a life for the 'other'.[37]

In order to do this, our education system and institutions within it must work towards:

a. Establishing knowledge as an intrinsic part of democratic politics

If, as we now know, our formal institutions of learning have been functioning using cultural canons that are culturally premised on western systems, a factor which has led to denigration by inertia, subjugation in practice and exploitation through research, then the IKS compels us to challenge the very *knowledge generation* and *legitimation processes* and become overt about the extent to which the knowledge we are producing reflects the true diversity of knowledge heritages. We need therefore to explore deeper the interface between epistemology, diversity and democracy, and the potential for true exchange and what Hountondji refers to as the 'reciprocal valorisation among knowledge systems'.[38] The intention here

37 Latouche, S. 1991, *In the Wake of the Affluent Society: An Exploration of Post Development*, Zed Books: London, p.17.

38 Hountondji, P. 1997, 'Endogenous knowledge. Research Trails', Codesria: Dakar, pp.13.

would be to establish knowledge as an intrinsic part of democratic politics.[39]

b. **Opening new moral and cognitive spaces**
This will enable us to *move the frontiers of discourse* and understanding in the sciences as a whole and to *open new moral and cognitive spaces* within which constructive dialogue and engagement for sustainable development can begin. In effect, it makes it possible for us to 'clear space' in order to *enable new issues* in science development to be generated and fostered, and thus determine new directions for the philosophy and sociology, as well as political economy of the sciences.[40]

c. **Understanding the political economy of 'Othering'**
Finally, we need to develop a clearer sense of the ethical and juridicial domains within which science works, and especially to begin to understand the political economy of 'Othering'. More importantly, the IKS humanises our practice, and enables us to become part of an empowering process for those silent witnesses of marginalisation (i.e. those regarded as *refractory to the scientific gaze*) and strengthen their capacity to take an active part in questioning the competence and ethics of the professional expert.[41]

In a final word to Gould and Lever in their uniqueness and excellence: both science and religion represent attempts by humans to gain mastery over the environment. Both have left trails of death, and also promised life. The sins and possibilities of salvation are strongly embedded in both. Only the implementation strategies are different. I propose that we commit to refrain from the death-giving, and to reinforce the life-enhancing essences of both, in concert with other traditions of knowledge, faith, and cultural systems likewise oriented.

39 Visvanathan, S. 1997, *A Carnival for Science: Essays on Science, Technology and Development*, Oxford University Press: Calcutta, p.2
40 Visvanathan, S. 1997, *A Carnival for Science*, pp.7–8.
41 Visvanathan, S. 1997, *A Carnival for Science*, pp.9–13.

Islamic Discourse on Evolution: Response to Science, Evolution and Schooling in South Africa by Jeffrey Lever

Abdulkader Tayob

Jeff Lever should be commended for presenting a paper that interrogates science and science education in broader society. By raising some critical questions about theology and science, he has challenged the smugness of religious institutions in the country. He is forcing these institutions to deal squarely with scientific developments and discoveries. Taken in its broadest sense, moreover, he has also challenged the notion prevailing in sectors of higher education and government that science and mathematics education has no organic link with beliefs, values and social development. The close link between society and science requires such perceptive and critical insights.

For a Muslim perspective on the questions raised, this paper challenges respondents to fill in the gaps, the missing links, in his paper. He presented a history of evolution and Christian theology and education; now the question naturally turns to other religious traditions in the country. Can one fill in the gaps in this paper by pointing to similar developments in Islamic education and theology?

A more general, critical reflection of the paper highlights deeper issues about science and religion in South Africa. Does the framework of Lever fit Islamic education or black education? Are there important areas that the paper overlooks? I will first provide some ideas that complement Lever's paper, and then offer a few critical comments.

A quick and convenient Muslim response to Lever's stimulating paper on evolution in South African education does not need much elaboration. It would be simply a modified affirmation of what Lever suspects is true of Islamic attitudes to the question:

> Just how strong a fundamentalist Islam is in the country is unclear. But those inclined that way are also vehement in their rejection of Darwinian evolution. The latter have adopted many of the stock arguments of their Christian counterparts, as any simple search on the Internet will demonstrate.[1]

This remark is situated within Lever's discussion of charismatic Christian objections to evolution since the 1970s.

The leading intellectual counterpart of fundamentalist Islam would be Islamisation of the sciences. According to this intellectual trend, Muslim scientists are called upon to restructure modern sciences, social and natural, on the basis of Islamic values: 'The reconstruction requires an original redefining of the *ummah's* intellectual and socio-cultural potentialities on the basis of its values, principles and historical perspectives, while keeping its doors open to the suitable contribution of others'.[2] Islamic misgivings and doubts about evolution have been conceptualised under this framework.

A number of private Muslim schools in South Africa adopted Islamisation as a guiding principle. The first such school was founded in 1985 in Cape Town by parents who feared that school boycotts in coloured areas would disrupt their children's

1 Lever, J. 2002, *Science, Evolution and Schooling in South Africa,* Africa Human Genome Initiative: Cape Town, p.24.

2 Basha, A.F. 1992, 'Islam, modern scientific discourse and cultural modernity: The politics of Islamisation of knowledge as a claim to de-westernisation', in *American Journal of Islamic Social Sciences*, vol. 2, no. 2, Summer.

education. Since 1994, however, many more schools have mushroomed all over the country, as middle-class Muslims fear the change and transformation occurring in the new South Africa. In such schools, I suspect that doubts towards evolution are framed under the broad rubric of Islamisation.

Some definitions of Islamisation may sound like the Doppers or the more recent charismatic Churches, but others sound decidedly like Christian liberals. Islamisation can sometimes look very much like the response of Johannes du Plessis of Stellenbosch (as outlined in the paper by Lever). Rejecting the literalism inherent in most understandings of Islamisation, many of its leading intellectuals propose a more moderate convergence between scripture and science. The following definition of Islamisation is instructive:

> Islamisation is the liberation of man first from magical, mythological, animistic and national-cultural tradition, and then from secular control over his reason and language.[3]

Syed Muhammad Naquib al-Attas has founded an institute in Malaysia committed to exploring the philosophy of Islam in all sciences. His definition of Islamisation is ambitious as it tries to locate an overarching narrative for Islamic sciences. More directly, Isma'il al Faruqi (d. 1986) inspired many South African Muslims to pursue science under the framework of Islam as a liberating force:

> Through *tawhid* [unity of God], therefore, nature was separated from the gods and spirits of primitive religion. *Tawhid* for the first time made it possible for the religio-mythopoeic mind to outgrow itself, for the sciences of nature and civilisation to develop with the blessing of a religious worldview that renounced once and for all any association of the sacred with nature. *Tawhid* is the opposite of superstition or myth, the enemies of natural science and civilisation.[4]

In both quotations above, one witnesses a careful openness to sciences from a modernist perspective to religion. In both cases, however, there is also a desire to create or preserve a religio-cultural framework for science. Many Muslim intellectuals fail to

3 al-Attas, S.M.N. 1985, *Islam, Secularism and the Philosophy of the Future*, Mansell Publishing Limited: London & New York, p.39.

4 al Faruqi, I. & al Faruqi, L.L. 1986, *The Cultural Atlas of Islam*, Macmillan: London & New York, p.80.

sufficiently clarify the extent to which scientific methods would or could encroach upon each other. They seem more concerned about affirming a holistic framework for science. Further research may be helpful in highlighting which Islamic trends, schools and theologians in South Africa support a literalist approach to Islamisation and which support a reformed modernist approach. More directly, research could interrogate the effect of Islamisation on the development of science in society. A study in Pakistan certainly points to the negative effect of Islamisation on science education.[5] But the varying definitions of Islamisation, from the literalist conservative to reformist, preclude us from a preconceived judgment in a country such as South Africa. It seems that Islamisation in South Africa is more a badge of identity than a deep philosophy of life.

But such a filler to Jeffrey Lever's insightful paper is insufficient. The paper raises broader issues that call for more thoughtful responses than simply filling in the gaps of his Islamic materials. The paper prompts at least four additional points worthy of discussion:

1. It must be pointed out that the title of the book is a misnomer. Purporting to read about 'Science, evolution and schooling in South Africa', one is immediately struck by its parochial nature. The discussion is entirely restricted to white educational experience of evolution and Darwinism in South Africa. The Islamic responses cited above are the only revelation that South Africa has more than one educational tradition, and a minuscule one at that. Even then, Muslim responses are presented as a reflection of Christian fundamentalism. The contexts and debates are completely absent, and the borrowed relationship with Christianity is quite misplaced as far as the power and location of Islamic educational experience is concerned.

   A brief review of Islamisation of education in the 1980s is indicative of the motives and location for this approach in South Africa. But apart from the Islamic experiences, education

5 Hoodbhoy, P. 1991, Preface in Trans. Professor Abdus Salam, *Islam and Science: Religious Orthodoxy and the Battle for Rationality*, Zed Books: London.

in all its rich, multiple and sometimes confusing history is obliterated. A framework is presented of religion and evolution through the prism of Christian theological developments. Within that clear and uncluttered narrative, Muslim, Jewish, Hindu, black, coloured, Indian, missionary, non-racial and multi-racial educational experiences must somehow find their place. This approach of presenting a framework of white experiences into which everybody else has to fit him- or herself, is a persistent experience of post-apartheid South Africa.

2. It is illuminating to read the intellectual and theological struggles between the many white antagonists and protagonists of evolution. But the choice of Jan C. Smuts among the latter is surprising, though nevertheless revealing, in regard to this epic struggle between saints and villains. While Lever cites Smuts's laudatory remarks on Darwin, he omits Smuts's appeal to social Darwinism in his Rhodes Memorial lecture in 1929. The African, he says, 'has largely remained a child type, with a child psychology and outlook. A child-like human can not be a bad human, for are we not in spiritual matters bidden to be like children? Perhaps as a direct result of this temperament the African is the only happy human I have come across'.[6]

   Judging from this example, it seems that the lines in South African white discourse between scientific evolutionists and social Darwinists may not be as clear as in Lever's presentation. The abuse of science in colonial contexts deserved more than a recovery of some brave souls who saved the honour of the scientific enterprise. A deeper structural analysis of science in South Africa as a whole needs attention.

3. The above leads to a third and crucial point about the need for a framework or narrative to help students understand biology (and indeed all other sciences) by using evolution as an overarching framework. The evolutionary framework is useful; it certainly helped me to study and understand Biology at university. Even though I believed in the literal truth of the Qu'ran then, I don't recall feeling a major existentialist angst in the face of

6 Mamdani, M. 1996, *Citizen and Subject: Contemporary Africa and the Legacy of late Colonialism*, Fountain Publishers: Kampala; David Philip: Cape Town; James Currey: London, p.4.

evolutionary scientific truth. At some higher level, I believe, things would be resolved. Now, however, I fail to see how this framework is absolutely essential, as Lever seems to think.

Lever's paper presents evolution as a theory, but one that will be difficult if not impossible to replace. It has stood the test of time, and can unlock the secrets of the universe and of the human condition itself. Such a narrative seems too dogmatic for science. The crucial role of understanding biology is certainly clear from the book, but Lever also suggests that evolution can unlock the deep nature of human life as such. Clearly, Lever does not propose that we recall the evolutionary bias for the social sciences as presented in the work of Weber, Durkheim, Freud and Comte. But it seems that the evolutionary key to unlock the meaning of human life and the universe remains unclear and perhaps unsubstantiated.

4. Against the need for an overarching framework, we should focus more on the scientific method as the key to open all doors. Narratives and frameworks are important for the educational experience, but the revolution of science lies in its method. One narrative, however plausible and defendable it may be in science, will be received as the dogma of a new order. The pluralistic history and experience of South Africa is much too precious and rich to obliterate with one framework. Scientific method should stand at the forefront of our struggle to improve scientific education and discovery in South Africa. Religious misgivings about scientific theories may be better debated with a general understanding of method than with an overarching narrative.

In conclusion, then, the discussion on evolution and science has provided an ideal opportunity for us to reflect on the interface between science and religious beliefs, and indeed between science and social attitudes. The particular case of Islam suggests that a contextual and critical reflection of Islamisation indicates that Islamic fundamentalism should not be compared with Christian fundamentalism. This does not mean that Islamisation does not pose methodological obstacles to scientific development, but Lever's book does not lay to rest the lines of demarcation between good science, politics and religion. While the need for evolution as a framework for biology is indisputable, evolution's promises to open the key to all mysteries is less convincing. Together with a revisable framework, we need a firm foundation for science on method.

Evolution, Creationism, Indigenous Knowledge: A Comment From a Somewhat Dissident Jewish Perspective

Denis Davis

For the orthodox Jew, the words of the Torah represent a set of eternal truths which are valid for all time. The question which has therefore vexed Jewish scholars over time is the extent to which these eternal truths, if valid for *all* time, should be related to the broader body of human knowledge.

There would appear to be no irreconcilable conflict between science and the basic tenets of Jewish faith insofar as the age and evolutionary development of the earth and the universe are concerned. Furthermore, the story of creation, particularly as set out in the *Book of Genesis*, can be interpreted so as not to be in conflict with the established age of the earth and the many billions of years more of the universe. Similar arguments can be presented for the general outline of the theory of evolution.

The pre-eminent commentator on the Torah, Rashi, does not provide a literal interpretation of the first chapters of the Book of Genesis. Rather, he interprets the first words of the Book in the following way:

> In the beginning of God's creation of heaven and earth, when the earth was without form and void, and darkness was on the face of the deep, and the spirit of God hovered over the face of the waters, God said 'Let there be light' and there was light.

The Bible does not tell us, says Rashi, that the order of creation was first heaven and then earth and then light. If that were the case Rashi suggests that the Bible would have said *Barishonah Bara Elokim et Ha'shamyim Ve'et ha'aretz* (first God created heaven and earth). The word *bereshit* (instead of *barishonah*) is always attached to words after it. Accordingly the phrase 'in the beginning God created' should rather be read as 'in the beginning of God's creation of'. Judah Landa[1] suggests that comments, interpreted this way, indicate that the Bible does not tell us that heaven and earth were created on the first day. The only creation to occur on the first day is that of light. The Bible merely puts the creation of light in context – that it occurred at the beginning of the process of God's creation of heaven and earth (perhaps meaning the creation of all that exists) when the earth was still unformed and void, 'darkness was upon the face of the deep, and the spirit of God hovered over the face of the water'.

When the Hebrew words of the text are examined within the context of the biblical story of creation, it does not mean that when we begin the narrative all that existed was created from nothing, but rather that the earth, lightness and dark that existed before the first day were fashioned to create this universe.

An illustration of this point is contained in the *Midrash*, where it reads: 'It doesn't say "let there be evening, let there be morning", but "there was evening, there was morning" '. From this we learn that evenings and mornings came and went before this day.

The first Chief Rabbi of Palestine, Rabbi Kook, wrote: 'It is absolutely immaterial to us whether there was in fact in antiquity a golden age when man enjoyed an abundance of material and spiritual wealth, or whether life began from the bottom and rose from the lowest rung of existence towards the higher, and that it continues upwards. What we must recognise is that it is distinctly

2 Landa, J. 1991, *About Torah and Science*, KTAV Publishing House, p.314.

possible that man, even after he has risen high, can forfeit everything by wickedness, and that he is likely to harm himself and generations to come. This is what we should learn from the story of Adam in paradise.'

Rabbi Kook therefore rejected the literal exegesis of the second chapter of Genesis and maintained that the chapter describing creation required a profoundly mystical interpretation. While many of his contemporaries regarded the concept of evolution as a threat to religion, Rabbi Kook considered evolution congenial to the deeper insights of Jewish mysticism, which had always viewed the world as continuously evolving towards the goal of ultimate perfection. He maintained that there was a passionate purpose and direction in evolution: the overwhelming longing of the human being was to cleave to God, which could only be attained through the progressive efforts of generations, each moving closer to the goal of holiness.

From the textual analysis of the Book of Genesis (read through the Hebrew), together with the insights derived from Rabbi Kook, a significant implication can be drawn for the treatment of evolution in schooling within South Africa as outlined by Jeffrey Lever in his paper. There is a respectable body of Jewish thought which does not see the Bible in literal terms; it sees the Bible in the terms outlined by Kook, namely there truly is no beginning and no end. Life is a process. There can be no Utopia because, if such a state were obtained, it would be compromised by the degrading realities of life. Thus a theory of evolution is easily compatible with foundational Jewish belief. Rabbi Kook put it best when he said, 'Everyone knows that the creation is one of the mysteries of the Torah and if all statements are merely to be taken literally, what mysteries are there?'

My contention is that the Torah needs to be read diachronically. A literal textual approach should be rejected, and in this fashion the Torah becomes less a set of biblical stories and more a broader framework for understanding what God expects of human beings, rather than the manner in which human beings analyse what God is supposed to have done.

It is no accident that biblical literalism accompanies religious fundamentalism. There is only one text, there is only one interpretation, there is only one answer. All difference is discounted and rejected. The 'other' becomes the enemy of faith and hence of God and should be purged from society. The politics relating to evolution in the classroom concerns a struggle which is far more important than the veracity of a particular scientific theory. It relates to fundamental questions of human freedom, of deliberation and ultimately of democratic society. The reason that fundamentalists are so vehement in their rejection of Darwinian evolution is that it destroys their certainty in the text as well as their ability to ensure conformation to one simple, clear and understandable interpretation. Even a cursory examination of the literature of Jewish sages will reveal its inherent pluralism in a world outlook, particularly in the spheres of cosmology and biology. Viewed in this context, this form of pluralism is critical to schooling in the kind of society prefigured in our constitution.

Christianity and Evolution

David Chidester

If the letters page of a daily newspaper can be a barometer of public concern, the opinion section of the *Cape Times* recorded particularly stormy weather during the winter of 2002 in the controversy over evolution. Following media coverage of Jeffrey Lever's critique of the failure of South African public schools to teach evolution, letters to the editor flooded in. After weeks of heated arguments, dogmatic assertions, desperate appeals to evidence and thinly veiled insults on the topics of science and religion, the editor ruled, 'This correspondence is now closed'. Still, the thunder and lightning continued. What was going on?

As the letters to the *Cape Times* suggested, the intellectual debate over evolution, such as it is, appears to have reached an impasse. Positions have already been staked out. Non-theistic evolutionists are pitted against theistic creationists, irreconcilably, with theistic evolutionists trying to play a mediating role by proposing that evolution might well be the means by which God creates. While committed evolutionists and creationists appear to hold each other in contempt, talking past each other at every turn, they at least seem to agree that they do not need or want any help from mediators who would try to distil a compromise position out of their opposition.

Perhaps an inevitable outcome of such a polarised controversy, this rejection of mediation is unfortunate because the most interesting, creative activity in both science and religion arises not from accepting or rejecting a proposition, but from renegotiating the terms in which that proposition is formulated. In some respects, the educational mission to advance teaching and learning about evolution in schools recalls the 19th century Christian mission in South Africa. Confronted with the Christian gospel, Africans accepted or rejected its appeal, but they also renegotiated its religious terms.

In mid-19th century colonial Natal, for example, these options can be briefly illustrated. Accepting Christianity, the convert, catechist and eventually deacon Mpengula Mbande completely embraced the religion and science of the Christian mission, arguing that Africans had been ignorant when they placed their trust in their ancestors rather than in Christ, and also proposing that European Christians held superior knowledge that might be regarded as scientific. Although all human beings emerged from the same source, the *Uhlanga* – the primordial bed of reeds from which all life originated – Mbande argued that Europeans had remained longer than Africans in the *Uhlanga*, drawing out all the wisdom, which was most evident in the technology of ploughs, wagons and firearms, so that when they came into the world they were able to conquer not by force but by superior knowledge that empowered them, as Mbande put it, to be 'victorious by sitting still'.[1]

By contrast, the convert William Ngidi, assistant to the bishop, missionary and biblical scholar John William Colenso, ultimately rejected the Christian mission because he found it inconsistent with indigenous African religion, especially with regard to Christian proscriptions against lobola and polygamy, but he also found parts of the Christian Bible contrary to science. While reading the biblical account of Noah's Ark, Ngidi demanded, 'Is all

1 Callaway, H. 1868–70, *The Religious System of the Amazulu*, Springvale Mission Press: Springvale, South Africa, pp.79–80; Chidester, D. 1996, *Savage Systems: Colonialism and Comparative Religion in Southern Africa*, University Press of Virginia: Charlottesville. pp.166–67.

that true? Do you really believe that all this happened thus?'[2] Based on his own observations of nature, Ngidi concluded that the story of Noah could not possibly have been a scientific fact, a position that Colenso also adopted in his biblical scholarship. While Colenso was charged with heresy, William Ngidi retreated to rural Natal and attempted to revive the ancestral traditions of African religion.[3]

Between total acceptance and total rejection, however, Africans found considerable room to manoeuvre. In response to Mbande's acceptance of Christianity, one Zulu by the name of Usithlanu qualified his account of the origin of European Christian religion, science and power by insisting that it had all been drawn from 'our' Zulu primordial source (*Uhlanga*), or 'our' Zulu ancestral origin (*Itongo*), suggesting, at the very least, that Europeans had no privileged ownership of powerful knowledge.[4] At the same time, in relation to Ngidi's rejection of Christianity, an African theologian such as Ira Nembula in the 1870s could tell African traditionalists that as Christians they 'held the faith of their own people', proposing that Christianity confirmed rather than contradicted the basic religious convictions of indigenous African religion.[5]

These examples illustrate the complexity of mediation between the polarised extremes of acceptance and rejection. Africans in 19th-century South Africa faced the challenge in different ways. So did European and Euro-American Christians. During the 1870s in Belfast, the prominent theologian Robert Watts rejected evolution, Henry Calderwood in Edinburgh embraced evolution, and Benjamin B. Warfield in Princeton adopted a mediating position by tolerating evolution because he separated science, for which he showed a keen interest, from faith in the Christian gospel of sin

2 Colenso, J.W. 1862–1879, *The Pentateuch and Book of Joshua Critically Examined.* 7 vols. Longman, Robert and Green: London, vol.1: p.vii.

3 Etherington, N. 1978, *Preachers, Peasants, and Politics in Southeast Africa, 1835–1880: African Christian Communities in Natal, Pondoland and Zululand,* Royal Historical Society: London, pp.43, 135, 158.

4 Callaway, H. 1868–70, *Religious System of the Amazulu,* Springvale Mission Press: Springvale, South Africa, p.94; Chidester, D. 1996, *Savage Systems,* p.167.

5 Etherington, N. 1978, *Preachers, Peasants, and Politics in Southeast Africa, 1835–1880,* pp.145–46; Chidester, D. 1992, *Religions of South Africa,* Routledge: London, pp.59–61.

and salvation.[6] Although they were all Calvinist Christians, these religious leaders advanced different strategies of engagement with the challenge posed by evolution.

If we were interested only in the interplay of science and religion as ways of knowing, we might explore further the possibility that everyone is always involved in processes of mediation because there are actually no 'pure' positions that can be established. Part of the messiness certainly results from the entanglement of knowledge with power. If knowledge is power, as Mpengula Mbande suggested, then knowledge is inevitably mixed up with political interests. In South Africa, as Jeffrey Lever recalls, the controversy over evolution has certainly been tangled up with the politics of religion and a religious politics. Under the aegis of Christian National Education, the apartheid regime – a self-proclaimed Christian state – explicitly excluded the teaching of evolution from the school curriculum on religious grounds. But Christian educators invoked religion not merely in response to challenges posed by certain kinds of scientific knowledge, but also in support of a specific kind of Christian political project.

In a widely used manual for Religious Education, published in 1983, and still in print during the 1990s, the authors advanced an argument against evolution that was political rather than scientific. 'Evolutionism is the intellectual basis of some of the anti-Christian and anti-God systems produced by godless men and their society', they asserted. 'The evil fruits of these systems is evidence enough of its opposition to God and his revelation'.[7] In retrospect, it is surprising to see such a consequentialist ethical argument, asserting that 'evil fruits' are sufficient evidence to indict a political system as evil, pronounced so glibly by Christian educators advancing the religious underpinnings of apartheid South Africa. Nevertheless, this political argument suggests that the controversy over evolution engages a complex set of interests at the intersection of science, religion and politics.

6 Livingstone, D.N. 1999, 'Situating Evangelical Responses to Evolution', in Livingstone, D., Hart, G. & Noll, M. (eds.) *Evangelicals and Science in Historical Perspective*, Oxford University Press: New York and Oxford. pp.200–12.

7 Kitshoff, M.C. & van Wyk, W.B. 1983, *Method of Religious Education and Biblical Studies*, Maskew Miller Longman: Cape Town. p.170.

Although Jeffrey Lever demonstrates that opposition to evolution, both religious and political, was mounted by the apartheid regime, he does not go sufficiently far in examining the effects of transnational religious organisations, mostly sourced from the United States, in advancing the anti-evolution crusade. During the mid-1980s, when the Dutch Reformed Church was wavering in its unconditional religious support for apartheid, the apartheid regime looked to conservative Christian organisations in the United States, the so-called New Religious Right, to provide ideological legitimation for its Christian crusade against communism.

This religious exchange has been extensive and sustained. We are still living with the consequences. In the field of education, for example, new initiatives in teaching and learning about religions, and religious diversity in our schools have met with vocal opposition from Christian groups in South Africa with links and funding from right-wing Christian organisations in the United States.[8]

In the controversy over evolution, most of the arguments against teaching and learning about evolution in schools seem to be coming from the same source: the operational catalogue of objections produced by conservative Christians in the United States. Having failed to establish any credibility for 'Creation Science' as an alternative theory for a science curriculum, right-wing Christian organisations have more recently mobilised around a new scientific project, 'Intelligent Design', which has been described by one commentator, unkindly perhaps, as nothing more than 'Creationism in a cheap tuxedo'.[9] For all their interest in science, in countering the teaching of evolution these projects are

8 Chidester, D. 2002, 'Religion Education: Learning about Religion, Religions, and Religious Diversity', in Asmal, K., & James, W. (eds.), *Spirit of the Nation: Reflections on South Africa's Educational Ethos*, New Africa Education: Cape Town, pp.93–104.

9 Melott, A.L. 'Intelligent Design is Creationism in a Cheap Tuxedo', *Physics Today* 55, 6 (http://www.physicstoday.org/vol-55/iss-6/p48a.html; accessed on 10 July 2002). For an account of earlier controversies over Creation Science, which is still promoted by conservative Christian organisations, such as the Institute for Creation Research and the Creation Research Society, see Numbers, R.L. 1992, *The Creationists*, Knopf Publishers: New York. Sponsored by a conservative think-tank, the Discovery Institute, prominent advocates of Intelligent Design include Johnson, P.

driven by a Christian political agenda, formulated most bluntly by Christian Reconstructionists, as capturing the curriculum of schools as a prelude to capturing control of the country.[10]

For ordinary Christians trying to work out the implications of all this for their own lives, as well as for the well-being of their children, the adversarial, combative politics of right-wing Christian organisations has only further polarised matters. If Jeffrey Lever finds that South Africa, with its legacy of Christian, apartheid opposition to evolution, is suffering from scientific illiteracy, we cannot view ourselves as so far behind the global curve if we refer to polling data from the US over the past 20 years, and learn that its population has been consistently divided between the 50 per cent who accept evolution and the 50 per cent who reject evolution. Still, we cannot feel good about the polarisation.

Although right-wing, conservative religious claims about science have been debilitating, scientists who have recently been proclaiming about religion are also not helping. Non-theistic evolutionists might generally dismiss religion as irrelevant in the quest for legitimate scientific knowledge, but evolutionary biologist Richard Dawkins has gone further by entering the field of religious commentary, asserting his personal conviction that 'Darwin made it possible to be an intellectually fulfilled atheist', but also publicly attacking religion as essentially dangerous – a 'misguided missile' that produces not only ignorance but also the kind of destruction witnessed on September 11, 2001 in the US. In the face of such aggressive scientific intervention in the field of religion, a mediating proposal such as Stephen Jay Gould's division of labour between two non-overlapping *magisteria* – religious faith and scientific knowledge, one charged with the

9 (continued)
1993, *Darwin on Trial.* InterVarsity Press: Downers Grove, Ill.; Behe, M. 1996, *Darwin's Black Box: The Biochemical Challenge to Evolution*, Free Press: New York; and Dembski, W. 1999, *Intelligent Design: The Bridge Between Science and Theology*, InterVarsity Press: Downers Grove, Ill. For critical commentary on Intelligent Design, see Pennock, R. T. 1999, *Tower of Babel: The Evidence against the New Creationism*, MIT Press: Cambridge, Mass.; and Pennock, R.T. (ed.) 2001, *Intelligent Design Creationism and Its Critics: Philosophical, Theological, and Scientific Perspectives*, MIT Press: Cambridge, Mass.

10 Chidester, D. 2002, 'Religion Education', p.97–98.

responsibility of morality, the other with knowledge about the natural world – might appear at first glance a welcome compromise.[11] However, for a religious tradition committed to the Augustinian imperative of *fides quaerens intellectum* – faith seeking understanding – this division of labour might also seem like a scientific dismissal of religion.

Scientists who have tried to fuse evolutionary science with their personal religious commitments, pursuing a *Scientist's Search for Common Ground between God and Evolution*, as Kenneth R. Miller subtitled his recent book, *Finding Darwin's God*, have developed forms of theistic evolutionism that tend to put God in the gaps in scientific knowledge rather than at the centre of a moral economy of sin and salvation.[12] Some theistic evolutionists have tried to move immediately to resolution, finding Darwinian grounds for Christianity or Christian grounds for Darwinian evolution, while others, seeking to establish an integration between science and religion, have deferred the realisation of that ultimate harmony into the indefinite future, in the meantime advocating ongoing dialogue between practising scientists and theologians.[13]

Scientists, however, are only asking for trouble when they use highly charged religious symbols in their rhetoric, employing ultimately religious language, not in scientific education, perhaps, but in attempts to popularise science for a broader audience. For example, on June 26, 2000, when the sequencing of the human genome was announced, US President Bill Clinton declared, 'Today, we are learning the language in which God created life'. While an American politician might be predisposed to intoning such religious rhetoric in seeking broad national

11 Gould, S.J. 1999, *Rocks of Ages: Science and Religion in the Fullness of Life*, Vintage: London; Ballantine: New York.

12 Miller, K.R. 1999, *Finding Darwin's God: A Scientist's Search for Common Ground between God and Evolution*, Cliff Street Books: New York.

13 Issues in current debates about science and religion are clearly set out in Barbour, I.G. 2000, *When Science Meets Religion*, HarperSanFrancisco: San Francisco. For different versions of theistic evolution, see Haught, J.F. 2001, *God After Darwin: A Theology of Evolution*, Westview Press: Boulder, Colo.; Peacocke, A. 1993, *Theology for a Scientific Age*, Fortress Press: Minneapolis, Minn.; Polkinghorne, J.C. 1999, *Belief in God in an Age of Science*, Yale University Press: New Haven; Ruse, M. 2001, *Can a Darwinian Be a Christian: The Relationship between Science and Religion*, Cambridge University Press: Cambridge; Wilson, D.S. 2002, *Darwin's Cathedral: Evolution, Religion, and the Nature of Society*, University of Chicago Press: Chicago.

support, a scientist is certainly courting religious controversy by invoking God in a discussion about a scientific discovery. Francis Collins, Director of the Human Genome Project and personally a devout Christian, entered such contested public terrain by resorting to religious language, asserting, 'We have caught the first glimpse of our own instruction book, previously known only to God'.[14]

Learning God's language, reading God's book – these religious claims invite religious engagement not only but certainly also by conservative Christians with a commitment to biblical creationism. Rhetorically, politicians and scientists, as in the case of Clinton and Collins, blur any distinctions that might be made among the spheres of political power, scientific knowledge, and religious faith by asserting religious significance for scientific discovery.

Although Collins is an evangelical Christian, even non-theistic scientists committed to non-theistic evolution have drawn on religious myths and metaphors when attempting to communicate the significance of their scientific discoveries to the public. The entomologist, evolutionist and sociobiologist Edward O. Wilson, for example, described his discovery of evolution as an 'epiphany', an illuminating, transcendent realisation of a new world, a replacement of his southern Baptist religious upbringing, which showed him that in the place of religion 'science is a continuation on new and better tested ground to attain the same end', that 'science is religion liberated and writ large'.[15] Such recourse to religious language is not the sole preserve of the life sciences, as Stephen Hawking in his best-selling book *A Brief History of Time*, echoed Clinton and Collins by representing the goal of theoretical physics as 'the ultimate triumph of human reason – for then we should know the mind of God'.[16] Understandably, given this extravagant religious rhetoric, some critics have argued that

14 Davies, K. 2001, *The Sequence: Inside the Race for the Human Genome*, Weidenfeld & Nicolson: London. p.236.

15 Wilson, E.O. 1998, *Consilience: The Unity of Knowledge*, Knopf: New York. p.4, 6.

16 Hawking, S. 1988, *A Brief History of Time: From the Big Bang to Black Holes*, Bantam: New York. p.193.

science has *become* a religion.[17] If religion has stepped on science's turf, as Richard Dawkins complains, scientists have also crossed into religious territory by making such ultimate claims.[18]

Staking out turf and territory, which is always negotiated and contested, shifting and at risk, suggests that Stephen Jay Gould was not entirely wrong in proposing separate domains for science and religion. Of course, he had no International Court of Justice to keep the combatants separate, so border conflicts will continue. Nevertheless, as a purely pragmatic proposal, we might adopt an educational division of labour with respect to evolution in South African public schools. In science education, evolution is taught, as Jeffrey Lever argues, not merely as theory or as fact, but also as the basic story that provides coherence to the entire scientific enterprise. Science education requires the teaching of evolution. In religion education, as part of Life Orientation, the Humanities and the Social Sciences, evolution is considered in relation to religious and other belief systems that provide meaning and values.[19] Here the story of evolution can be fruitfully related to other stories of ultimate coherence. Religion education requires teaching and learning about creationist, evolutionary and other belief systems in ways that foster creative and critical inquiry into our different and differing understandings of human origins and destinies.

In this educational work, we are faced with the challenge of enabling learners to tolerate and perhaps even respect difference, thus avoiding the exchange of insults in the opinion section of our newspapers, while expanding their interpretative resources and sharpening their critical skills for engaging religious and other convictions. In both science education and religion education, our public schools have the opportunity to create spaces of learning that are free from the dogmatic censorship and indoctrination of

17 For treatments of science as a religion, see Nelkin, D. 2001, 'Less Selfish than Sacred? Genes and the Religious Impulse in Evolutionary Psychology', in Rose, H. & Rose, S. (eds.), *Alas, Poor Darwin: Arguments Against Evolutionary Psychology*, Vintage: London. p.14–27; Midgley, M. 1985, *Evolution as a Religion: Strange Hopes and Stranger Fears*, Methuen: London; 1982, *Science as Salvation: A Modern Myth and Its Meaning*, Routledge: London; and Noble, D.F. 1997, *The Religion of Technology: The Divinity of Man and the Spirit of Invention*, Alfred A. Knopf: New York.

18 Dawkins, R. 1998, 'When Religion Steps on Science's Turf', *Free Inquiry*, 18, 2, p.18–19.

19 For an introduction to religion education, see Chidester, D., Mitchell, G., Phiri, I.A. & Omar, A.R. 1994, *Religion in Public Education: Options for a New South Africa*, UCT Press: Cape Town.

the past. Although the letters pages of our daily newspapers might continue to register stormy weather, public education can create safe havens for science education, informed by evolutionary theory, and religion education, dedicated to teaching and learning about the diversity of religious and other deeply-felt convictions held by people in our diverse society.

Biology, Evolution, Curriculum Development and Publishing

3

The Structure of the Natural Sciences Learning Area Statement and Opportunities Within it for the Teaching and Learning of Evolution

Dev Isaac

Introduction

In January 2001, the Natural Sciences Working Group was established to streamline and strengthen Curriculum 2005 in respect of the Natural Sciences Learning Area in response to the recommendations made by the Review Committee and approved by the national Minister of Education. Originally the Working Group comprised six members.

In July 2001, the Working Group's draft document, referred to as the draft Learning Area Statement for the Natural Sciences, was released for public comment, together with Statements of other Learning Areas. From December 2001, the Working Group re-drafted its Learning Area Statement to incorporate public comments and to finalise the document. At this stage, the Working Group was broadened to include four new members. The finalised document is currently being printed. Together with the Statements for the other seven Learning Areas, and an overview document, it forms the National Curriculum Statement for Grades R to 9.

The structure of the natural sciences learning area statement

The Natural Sciences Learning Area Statement includes generic sections which are the same for all Learning Area Statements, as well as specific sections dealing with the Natural Sciences. The generic sections deal with, *inter alia* outcomes-based education, Time Allocations, Assessment and the kind of learner that is envisaged. The Natural Sciences-specific sections include descriptions of the Learning Outcomes, Assessment Standards and Core Knowledge and Concepts relevant to the Natural Sciences.

The Learning Outcomes

The seven Critical Outcomes and the five Developmental Outcomes identified and described in the original version of Curriculum 2005 have been retained in the National Curriculum Statement. The Natural Sciences Learning Area Statement addresses these Critical and Developmental Outcomes through the following three Learning Outcomes:

- Learning Outcome 1 (Scientific Investigations): *The learner will be able to act confidently on curiosity about natural phenomena, and to investigate relationships and solve problems in scientific, technological and environmental contexts.*
- Learning Outcome 2 (Constructing Science Knowledge): *The learner will know and be able to interpret and apply scientific, technological and environmental knowledge.*
- Learning Outcome 3 (Science, Society and the Environment): *The learner will be able to demonstrate an understanding of the inter-relationships between science and technology, society and the environment.*

The Assessment Standards

The Assessment Standards are ways in which learners demonstrate the achievement of the Natural Sciences Learning Outcomes. The table overleaf shows the ten Assessment Standards and the Learning Outcomes from which they originate.

Learning Outcome 1 (Scientific Investigations)	Learning Outcome 2 (Constructing Science Knowledge)	Learning Outcome 3 (Science, Society and the Environment)	
		Intermediate Phase	**Senior Phase**
Planning investigations	Recalling information when needed	Understanding science and technology in the context of history and indigenous knowledge	Understanding the impact of science and technology on the environment and on people's lives
Conducting investigations and collecting data	Categorising information to reduce complexity and look for patterns	Understanding science as a human endeavour in cultural contexts	Understanding sustainable use of the Earth's resources
Evaluating data and communicating findings	Interpreting information	Recognising bias in science and technology which impacts on people's lives	
	Applying knowledge to problems that are not taught explicitly		

It should be noted that, in the Foundation Phase, the three Assessment Standards are expressed, in terms appropriate to the phase as:

- plan;
- do; and
- review.

The Assessment Standards for Learning Outcome 3 are written differently for the Intermediate and Senior Phases. This was done so as to align the Intermediate Phase Assessment Standards with those of the Technology Learning Area, in anticipation of the possibility that, in some schools, Natural Sciences and Technology may be combined into one Learning Programme in the Intermediate Phase.

Core Knowledge and Concepts

The Natural Sciences Learning Area comprises a wide variety of fields of inquiry. Because of this diversity, it is not possible to list all science knowledge under one heading. Therefore, the fields which scientists study have been grouped into four strands or content areas viz. Life and Living, Energy and Change, Planet Earth and Beyond and Matter and Materials. Each of these strands has been refined into sub-strands, each of which is summarised by a general proposition or unifying statement. The table below reflects the four main strands, their sub-strands, and the unifying statements:

Strand	Sub-strand	Unifying Statement
Life and Living	Life Processes and Healthy Living	Living things, including humans and invisibly small organisms, can be understood in terms of life processes, functional units and systems.
	Interactions in Environments	Organisms in ecosystems are dependent for their survival on the presence of abiotic factors and their relationship with other organisms.
	Biodiversity, Change and Continuity	The huge diversity of forms of life can be understood in terms of a history of change in environments, and in characteristics of plants and animals throughout the world over millions of years.
Energy and Change	Energy Transfers and Systems	Energy is transferred through biological or physical systems from energy sources. With each transfer, some of the energy becomes less available for use, and therefore we need to know how to control energy transfers.
	Energy and Development in South Africa	Energy is available from a limited number of sources, and the sustainable development of countries in our region depends on the wise use of energy resources.

⇨

Strand	Sub-strand	Unifying Statement
Planet Earth and Beyond	Our Place in Space	Our planet is a small part of a vast solar system in an immense galaxy.
	Atmosphere and Weather	The atmosphere is a system which interacts with the land, lakes and oceans and which transfers energy and water from place to place.
	The Changing Earth	The Earth is composed of materials which are continually being changed by forces on and under its surface.
Matter and Materials	Properties and Uses of Materials	We can classify materials by their properties in order to establish types and patterns. Properties determine the selection of materials for particular uses.
	Structure, Reactions and Changes of Materials	We can modify materials in ways we choose, through our understanding of their sub-structure.

The Learning Outcomes are the operations which learners must be able to do according to a certain range of scientific knowledge. The Assessment Standards derived from the three outcomes define the *levels* at which learners operate within an outcome. The core knowledge and concepts are neither Learning Outcomes nor Assessment Standards. They define the *breadth* over which learners can operate at any level. In other words, the core knowledge and concepts serve as the vehicle for delivery of the outcomes. Furthermore, the core knowledge and concepts for the Intermediate and Senior Phases represent a notional 70 per cent of the time in a Phase's Learning Programme. The other 30 per cent of the time should be used to extend these minimum knowledge statements; alternatively, science content from contexts which are significant to the learners and the local community may be used. These contexts may be economic, environmental, social or health matters, for example.

New Priorities and Challenges

In the development of the Learning Area Statement for the Natural Sciences, that is, in the writing of the Learning Outcomes, Assessment Standards and the selection of the Core Knowledge and Concepts, the following new priorities were taken into account:

- the need to place investigations at the centre of all classroom activities;
- the development of scientific literacy in all sectors of the population;
- the 'big ideas' of all the major sub-disciplines of Science in a balanced way, with due cognisance given to conceptual progression;
- integration with other Learning Area Statements, notably that of technology;
- the promotion of the Core Knowledge and Concepts in such a way as to enable learners to develop the necessary skills to take these 'big ideas' of science further;
- the impact of science and technology on people's lives and on the environment;
- the promotion of human rights and issues of social justice;
- the inclusion of indigenous knowledge systems as a means of understanding human endeavour within cultural contexts; and
- the provision of opportunities for the participation of all learners, including those who may have been previously denied these opportunities on the basis of race, gender, poverty, physical and/or psychological disability, and those living with such challenges as HIV/AIDS.

The Learning Area Statement for the Natural Sciences is much more than a catalogue of facts that needs to be taught and learned; it is a Curriculum Statement organised around the priorities of a society in transition while at the same time taking into account that South African learners should be given every opportunity to become leaders in a global arena.

Opportunities for the teaching and learning of evolution

The Learning Area Statement for the Natural Sciences does provide opportunities for the teaching and learning of Evolution. The third sub-strand of the strand Life and Living, as shown in the table on page 113, deals with Biodiversity, Change and Continuity. The unifying statement for this sub-strand is as follows:

> The huge diversity of forms of life can be understood in terms of a history of change in environments and in characteristics of plants and animals throughout the world over millions of years.

Change in characteristics of plants and animals over long periods of time *is* evolution.

More specific references to evolution within this sub-strand of the document are described in the following:.

(a) *Sexual Reproduction* (Intermediate Phase): This provides opportunities to discuss similarities and differences between parents and offspring.

(b) *Variation of Offspring* (Senior Phase): This provides opportunities to discuss the shuffling of genetic material, why the genetic constitution of an individual organism may be different from its parents, and variation of individuals within a species (this is an area that may be discussed in depth in the FET schools curriculum).

(c) *Natural Selection* (Senior Phase): This provides opportunities to explain the concept of 'survival of the fittest', i.e. of the wide variety of individuals within a species, only those suited to the environment, in terms of their physical and physiological constitution, survive and pass on their characteristics.

(d) *Extinction* (Senior Phase): This allows for the discussion of accelerated natural selection and extinction of species which may be brought about by changes in the environment.

(e) *Classification* (Senior Phase): This allows for the discussion of similarities and differences due to common descent and divergence of species.

(f) *South Africa's Rich Fossil Record* (Intermediate Phase): This provides obvious opportunities for the discussion of evolution. It should be noted that *fossils* are also treated in the sub-strand The Changing Earth of the strand, The Planet Earth and Beyond (Intermediate Phase).

Still other opportunities exist outside the sub-strand Biodiversity, Change and Continuity, for the teaching of *Evolution*. For example, the sub-strand Interactions in the Environment, includes *Adaptation to the Environment* (Senior Phase). Also included in this sub-strand is the concept of *Competition* (Senior Phase), which plays an equally important role in natural selection.

The prejudice that some human groups are at a more advanced stage of evolution than others, is evident in the document through the following statements:

> Variations in human biological characteristics, such as skin colour, height and so on, have been used to categorise groups of people. These biological differences do not indicate differences in the innate abilities of the groups concerned. Therefore, such categorisation of groups by biological differences is neither scientifically valid or exact; it is a social construct.

Earlier, mention was made of the fact that the core knowledge and concepts listed represent a notional 70 per cent of the time and that the other 30 per cent of the time should be used to extend these minimum knowledge statements and/or to select content from contexts which are significant to the learners and the local community. This, then, provides further opportunity to take the topic of evolution further, in communities where teachers and learners choose to do so.

It should be noted that this Learning Area Statement (and the National Curriculum Statement as a whole), has been written for school learners from Grade R to Grade 9, i.e. for those learners in the GET band. Any assessment of it must take into account that the 'big ideas' mentioned in it are most likely to be extended in the schools curriculum of the FET phase, currently being drafted.

Conclusion

Those given the responsibility of developing the Learning Area Statement for the Natural Sciences had the challenging task of balancing the national priorities of a society in transition within a new democracy, with the demands of ensuring that major concepts of the various fields of science are not only included, but also mapped out within an internationally accepted framework of conceptual development.

Evolution, as a 'big idea' of science, has its place within this framework. However, for reasons already explained, it does not enjoy priority of purpose, nor is it seen as the prime organiser of all knowledge within the Life and Living component of the Natural Sciences Learning Area. It is hoped that the FET schools curriculum will extend discussion of this concept into the Human Genome Project.

The focus of evolution in the GET band is on *understanding* the principles of evolution rather than *accepting* them dogmatically. In this way, the curriculum opens the way for debate and discussion, an approach which is not limited to this topic. While such an approach is an integral part of good quality teaching and learning, it is even more important in the multi-cultural context of South African classrooms where the world-views of all learners need to be respected.

Science, Evolution and Book Publishing: A Publisher's Dilemma

Fathima Dada

Introduction

This paper will attempt to contextualise publishing, in particular educational book publishing, within its domains and constituencies. The challenges faced by publishers, with particular reference to evolution, science and religion, will be highlighted.

The first half of the paper will make some observations of a general nature, emanating from a reading of Jeffrey Lever's paper, while the second half will specifically address the (implied) questions embedded within the paper, regarding the role of publishers in the question of science, evolution and the curriculum.

I must confess that my line of argument leaves unanswered the question of what (to use dated terminology) is 'right' and 'wrong'. Children do, however, have to develop the skills to identify the criteria for making choices.

There are two traps in this debate. On the one hand, the state or publishers could make the decision to support one or other of the theories of creationism or evolution. In this case children will be presented with one view of the world. On the other hand, they could uphold liberal relativisim, where two 'good' stories are told, and children have to make their own choices and judgements. But children must develop their own value systems and to do this they may need to know which is the dominant view of the day. In a sense, this is partially what a Human Rights-based curriculum is about. They need to know their own rights and responsibilities but also, for example, that the actions of someone like Slobodan Milosovic, or the tenets of apartheid, are wrong.

This is a difficult but laudable task, and exactly what the new curriculum statement tries to address.

Religious science or scientific religion?

Lever's paper attests to the 'fact' that: 'Darwin's theory stands stubbornly uncontroverted.' He reminds us that the theory appears time and again and never ceases to amaze with its ability to be applied to many facets of life and the real world. His observation is correct, but let us replace the subject with religion or creationism. How easily and smoothly the semantics read:

> Creation theory stands stubbornly uncontroverted. Creation theory pops up time and again, never ceasing to amaze with its ability to be applied to many facets of life and the real world ...[1]

The question, of course, is 'who' claims 'what' is 'uncontroverted'.

The point is that as long as the fervent few keep science and religion in stubborn opposition, their domains will never cross and they will forever remain antagonists rather than partners. Should we not use the school curriculum to present the case of each theory, demonstrating to the young and eager minds of our children that knowledge itself is evolutionary and never fixed?

Another way to resolve this dilemma is to teach creationism as part of religion education and evolution as part of science, and to deliberately keep their domains separate. The two stories are not necessarily in competition with each other.

1 Lever, J. 2002, *Science, Evolution and Schooling in South Africa*, Africa Human Genome Initiative: Cape Town.

Lever makes the point that science is ever changing, yet history reminds us that scientists at given moments in time have felt the conviction to make fixed statements themselves. Witness the change in the views of the leadership of the South African Association for the Advancement of Science. Duerden, as its president, made the historic statement in 1921 that: 'In his hereditary endowments the white is far more gifted than the coloured, and must lead.'[2] Duerden was a leading scientist at the time and was certainly likely to be believed by many, even if not by all.

Less than five years later, the then President of the South African Association for the Advancement of Science, Smuts, publicly supported the views of Raymond Dart (who had labelled Duerden's conclusions as racist), which were in opposition to his predecessor, and proudly proclaimed Africa to be the cradle of humanity and not Asia. Having said this, it must also be noted that Smuts, to the very end of his life, had a contradictory and ambiguous view of race.

What new theory, whether religious or scientific, will be discredited in a 100 years' time, or will have evolved to a point beyond recognition of where it is today?

History and common experience suggests that:

- religion will be with us for as long as we want and need it, for it is believed to be linked to the human endeavor towards spiritual growth;
- science will take us forward both technologically and in terms of understanding life itself; and
- both science and religion, in their search for the ultimate truth, make it more and more complex for ordinary people to make sense of life and the universe.

Therefore the notion that the school curriculum should be a tool for propagating particular theories is a dated one. We should rather use the opportunity to provide our children with the knowledge and skills to analyse a variety of views and arguments, encouraging them to engage in debate analysis, and to seek more knowledge.

2 Duerden, J.E. 1921, 'Social anthropology in South Africa: problems of race and nationality' *South African Journal of Science*. vol.24, p.30.

As part of this approach we need to teach children that 'at this point in history' evolution and creationism are unequal. The overwhelming evidence is in favour of evolution; in large part the way the modern world works is underpinned by evolution. But there are other points of view, and knowing all of this, children should make their own choices, even if they settle for creationism.

The separation of state and church: A happy divorce, by all accounts?

The recent (and timeous) separation of church and state in South Africa has had a profound effect on the South African school curriculum, felt to be not a moment too soon by some and far too soon by others.

This development, which appears to have taken the Christian right by surprise, has sounded the death knell for Christian National Education. The idea is to educate the future citizens of this country using the constitution as a starting point. It is an attempt to make the final move away from Christian Nationalism, and the 'hyperfactualism'[3] of the past curricula towards a more 'balanced' knowledge and skills-based curriculum. Below are some of the critical outcomes intended to permeate the entire curriculum. They are based on the national constitution and world-wide benchmarking, as follows:

- Learners should be able to collect, analyse, organise and critically evaluate information.
- Learners should be able to use science and technology effectively and critically, taking responsibility for the environment and the health of others.
- Learners should be able to demonstrate an understanding of the world as a set of related systems by recognising that problem-solving contexts do not exist in isolation.

The application to science, (or religion, for that matter), would be to focus on *how* science is taught rather than only on *what* is taught. Lever makes a good and relevant recommendation here

3 Lever, J. 2002, *Science, Evolution and Schooling in South Africa.*

that it would be more effective to teach science (especially at an early age) in a historical narrative than in a plethora of unrelated facts. Just as children find appeal and beauty in the stories of the Old Testament, Torah, Qu'ran, Gita and others, so will they find appeal and interest in the 'story' of evolution – their own, original story. After all, we cannot deny the existence of fossils, so let's use whatever scientific evidence we have to pique the interest and encourage the advancement of learners.

On the face of it, this happy divorce of state and church appears to have resulted in an admirable and inoffensive set of world-class curriculum standards, yet it has met with strong opposition.

So, what will the publishers do in their textbooks? To answer this challenging question more fully, we will first have to delve into the political realm.

Can and should we now separate the state from the curriculum?

Curricula (and textbooks) reflect the mainstream thinking and ideologies of the time. They are thus also evolutionary in nature. Witness the South African curriculum history: from 'Christian and national', to 'factual, national, selective and divisive' and, most recently, to 'constitutional and rights-based'.

The previous regime, as we all know, carried the state ideology into its school curriculum. The ANC deserves credit for being truly modern in its approach of refraining from imposing a state ideology on the curriculum. The curriculum is developed from a progressive ideology – human rights or the constitution – making South Africa one of the few developing countries to do this. How do we defend this state of affairs so that subsequent politicians do not unravel it?

Memory can be an elusive attribute, and a case may therefore be made for keeping politics out of the curriculum. After all, the same political party that gave us Christian National Education, Bantu Education and segregation, has thrown in their lot with their former enemy, the ANC. Is this political evolution or is this simply a modern form of natural selection?

The evolution of curriculum and the evolution of publishing: infinite parallels or partners?

A brief distinction between the worlds of curriculum development and publishing is warranted here. The curriculum development process is initiated in South Africa by government, via the National Department of Education. Experts in various subjects and fields are employed from either within or outside the Department to conduct research, develop the curriculum, test it and implement it. The curriculum process is not and should not be static; rather it should evolve over time based on a combination of the country's priorities and general education trends.

Education book publishing is distinctive from curriculum development in that it is just one component of the implementation of the curriculum, albeit an important one. The publishing process is based on research into what teachers and learners need for the curriculum to be implemented successfully, and for learners themselves to achieve success. Market forces such as cost, taste, communities and their preferences with regard to content, education department requirements and a host of other issues, are taken into account.

In the same way that science and knowledge is not static, neither is the curriculum nor is book publishing. They have distinct as well as common developmental processes.

What is the role of the curriculum? Does the curriculum have a responsibility to society? Does a curriculum have obligations beyond the political?

In tracing the history of curriculum in SA, it is clear that a country's curriculum can have a profound effect on society, both in terms of shaping its citizens and bringing about progress and change. It is clear too, that curriculum reform is an ongoing and evolutionary process. With issues of such gravity vested in the curriculum, it is not surprising that the recent and radical changes in the curriculum (compared to CNE), have resulted in the evident opposition and emotion.

What is the publishers' role? What is the publishers' responsibility? Do they have obligations beyond the commercial?

In trying to understand some of the challenges facing publishers, let us examine their imperatives and choices:

- Publishers (and their authors) are essentially interpreters and packagers of curriculum, knowledge and information. They attempt to do this in the most interesting and accessible way for teachers and learners.
- Book publishing is a commercial undertaking. If the market doesn't like it, the book won't sell, so it shouldn't be published.
- What does the curriculum say? A veritable bible for most publishers, the national and provincial curricula standards are the starting points for textbook development.
- Is there space for innovation? The market is not homogenous; and publishers do have substantial space, especially in the recent South African curricula, for innovation and varied offerings.
- How should publishers respond to market forces? Their options include:
  - Multiple courses representing different views and approaches; offering the market choice.
  - 'Safe' courses, steering away from controversial debates.
  - Bold and groundbreaking courses capable of forever changing the face of textbooks.

It has been said that the state of the book publishing sector is a good measure of a country's state of democracy. If this is true then we have little to fear, as South Africa has a relatively healthy publishing industry, ruled by market forces. This has not always been the case, however. There was a time when the prevailing political idealogues had a profound influence on the few dominant players in the book-publishing industry, showering privilege on authors and publishers who danced to the tune of the political masters. Even today, we still find books of dubious political affiliation being used in schools.

Evolution and textbooks

Turning to the specific case of evolution in textbooks, let us examine some popular local Standard 9/Grade 11 Biology textbooks produced over the past two decades.[4]

4 For obvious reasons, the full titles and bibliographic references of these books cannot be given.

Book 1 (published in 1986)

This popular book plays safe by saying nothing of evolution and has proudly survived since 1986, with 22 printing impressions.

Book 2 (published in 1986)

This book has a substantive chapter entitled 'The continuity of life'. It introduces the cell, chromosomes and DNA, cell replication and reproduction, and passes on to a discussion of genetics proper, which is further dissected into gametes, monohybrid crosses, dominance, segregation and assortment, through sex differentiation to arrive at natural selection. The reference to evolution is as follows: 'Natural selection is the mechanism of what is now known as the theory of evolution.' The book includes a section called enrichment (not for exam purposes) that postulates the theory of evolution and explains natural selection. This is then contrasted with the theory of creationism. It goes on to say:

> '[Some scientists] also believe that although natural selection can, in theory, explain the gradual changes in evolution, *there is no evidence* that it has actually brought about *a major evolutionary change from a simple unicellular organism into a more complex multicellular organism.* ... Whichever view one takes is largely a matter of faith. Neither organic evolution nor creationism can be proved scientifically in the laboratory.'

It is worth noting that the two titles mentioned above were published in the same year, yet have extremely different approaches to the subject of evolution, probably fairly representing the views prevalent in the two different companies at the time.

Book 3 (published in 1996)

This textbook has a reasonably empirical and methodical approach, which tries to guide the learner through the examination material as swiftly and as painlessly as possible. After introducing 'Genetic Mechanisms', it discusses Natural Selection. The message is so brief that it is worth quoting in full:

> Natuurlike seleksie draai om variasie en is die seleksie deur die omgewing van individue met die mees voordelike erflikheidsvariasies, beide die wat alreeds verstaan en nuwe variasies wat ook al mag staan. Die natuur selekteer daardie

individue wat die beste kans op oorlewing in hul omgewing besit. Dit word gedoen deur die mees bruikbare kenmerke by 'n spesie te selekteer en toe te laat dat dit na daaropvolgende generasies oorgedra word.

(Natural selection revolves around variation and is the selection through the environment of individuals with the most favourable inherited variations, both those which already exist and those which have yet to come into being. Nature selects those individuals which have the best chance of surviving in their environment. This is done by the selection of the most usable features of a species and by passing these on to future generations through heredity.)

The text then mentions gene pools and practical application, but the danger is over. Evolution has been elided without being dismissed.

Book 4 (published in 1999)

This book presents an extraordinarily bold view of evolution, beginning the book with the subject in an unequivocal manner. The opening remark states: 'This chapter explores the theory of evolution which seeks to explain the diversity of life on Earth.' The book not only explains evolution and natural selection in great detail but takes readers through the major South African religions. It is an honest attempt to accommodate a variety of beliefs. Testimony to this is a question in the book:

> How did life on Earth begin? Was it a random chemical accident, or was it part of a plan of a higher being? What do you think? Discuss your ideas in class.

Conclusion

If there is a pattern emerging in the handling of evolution in textbooks, it is that progress is being made over time and through curriculum reform and political change. The point is that Darwin's theory seems to be here to stay, and seems almost to be asserting itself in our classrooms as a reflection of our political and ideological development. We have made progress and cannot turn back.

This is also an appropriate point to end, as it aptly reflects the awkward marriage between politics, curriculum, science, religion and textbooks. But a marriage it is and must remain, taking the view that it is a partnership that will grow with understanding, knowledge, progress and tolerance.

The challenge remains to curriculum developers, authors of textbooks and publishers, to balance the demands of scientific evidence, religious and societal beliefs, commercial forces and sound education principles.

Good curricula and good textbooks will:

- present learners with what is currently believed to be fact;
- sensitise learners to religious, cultural and environmental contexts within which the knowledge or evidence exists;
- develop critical and analytical tools in learners through which they will grow and make sense of their universe; and
- maintain the academic rigour and integrity of the various disciplines and subjects.

Challenges of Writing about Evolution in School Textbooks

Colleen Dawson

Introduction

Writing appropriate textbook materials which deal with topics related to evolution is a challenge similar to the challenges faced by classroom educators and by curriculum developers in dealing with what can be a difficult and sensitive topic.

Discussion at this conference and the papers presented here illustrate the broad learning framework within which teaching and learning about evolution must be situated. This includes learning areas within science, history, geography and cultural/religious studies. The focus of this paper is to narrow the framework down to the concepts with which a textbook writer or educator must deal in the classroom in order to build learners' understanding to a point where they can begin to comprehend, if not accept, the idea of evolution of all living species, including humans.

Textbook writers and other educators have the important task of helping learners from Grade 1 onwards to build an understanding of core scientific and other concepts that relate to evolution, such as change, diversity, variation, adaptation, genetic inheritance and geological time frames. This paper discusses how some of these concepts can be developed within a textbook framework at different school phases in order to ensure that learners gain the tools for understanding this fascinating topic.

Past curricula in South Africa

Jeff Lever has discussed the relevant issues in South African curricula from the 1920s to the present. For a textbook writer, a major challenge is that textbooks and curricula have not adequately dealt with evolution and related topics in the past. Educators have not themselves learned much about the related topics in their own schooling or professional training. Textbooks and the curricula from which they are developed play a crucial role in building up the necessary framework both for educators and learners.

Building up a conceptual framework for the teaching of evolution

School science (and geography and history) textbooks, curricula and teaching, all need to provide a comprehensive framework within which conceptual development – to build an understanding of the complex ideas of evolution – can take place. This needs to start in Grade 1 and continue through the GET phases into FET, to ensure that the concepts are taught early enough and appropriately enough for the level of the learners.

To do this, one of the first tasks of a science educator must be to help learners work with the scientific meaning of the words *theory* and *hypothesis* compared to their common usage. It is critical that learners are given the opportunity in their science lessons to develop hypotheses themselves, and to test them in order to confirm or reject these. If they confirm a hypothesis, they need to find another way of testing it, until there is evidence which leads to calling the original hypothesis a theory. Often learners continue to use the words interchangeably, and feel they can reject any learning that is called a theory because they see it as merely somebody's idea or hypothesis, rather than a thoroughly researched and tested body of knowledge which forms the best explanation available for observable phenomena.

Topics which can be dealt with in the different phases in order to build up the framework needed to deal with adaptation and change in the Senior Phase of GET, and into the FET Phase where evolutionary theory is dealt with more fully, are suggested here. This list is certainly not comprehensive, but is offered in order to illustrate that textbook writers and other educators need to be aware of how conceptual development in their topic is handled. The lists are offered also to emphasise how early the learning of Science must occur to build a scientific framework within which learners can view the world. Following these suggestions for the three GET phases one topic is selected from each phase, and a fuller explanation is presented, showing how it can be developed into a textbook unit which will stimulate an understanding of evolution. This is done within an OBE framework of teaching and learning skills, knowledge and values.

Some of these are mentioned in the recently released revised curriculum.

- **Suggested broad areas for Foundation Phase**
  - Change
  - Diversity of plants and animals in your environment
  - Variation within a species in your environment
  - Comparison of plant fossils with living organisms
  - Scientific method
- **Suggested broad areas for Intermediate Phase**
  - Adaptation of organisms with respect to an environment
  - Scientific method
  - Comparative structure and functions of systems in plants and animals
  - Geological time periods, patterns of life over time, fossil evidence, other evidence
- **Suggested broad areas for Senior Phase**
  - Darwin's experience and how he came to his theory of natural selection
  - Change – comparative reproduction and growth in plants and animals
  - Changing ecosystems – species interacting with each other

- Comparing and contrasting of the adaptation of organisms to conditions in biomes
- Fossil records
- Extinction

▪ **Suggested broad areas for FET Phase in schools**
  - Genetics
  - Adaptation through variation
  - Change
  - Adaptation to the environment
  - Comparative structure and functions of systems in plants and animals
  - Evidence for evolution – Darwin's research, distribution studies, comparative anatomy, palaeontology, cell and molecular biology

Development of topics into suitable textbook units

The important general concepts that need to be included within a Science course in order to build up understanding of major scientific ideas such as evolutionary change, include those of change, adaptation, diversity, variation, using the scientific method of investigation, comparative structure and functions of systems and organs in organisms, and genetics. All of these broad concepts can be dealt with at a simple level in the Foundation Phase and in much more complex ways at any higher level of schooling or further education.

Another language issue that arises from the list is the difference in science usage of *variation* and *diversity*. *Variation* means the differences between individual plants or animals that are from the same species. *Diversity* refers to the different species of plants or animals.

Below are suggestions for how a textbook writer might deal with one of the areas in a textbook unit for each phase, which a classroom educator could then use for the basis of a lesson or lessons.

Foundation Phase textbook unit

Learners in Grade 1 can begin to work with ideas of *change over time* using scientific skills and the scientific method. A suitable textbook unit or units would present activities involving learners observing fossils (or pictures of fossils in the textbook, if no other resources are available) and comparing them to present-day organisms. The fossils could include those of plants and animals.

This unit could be combined with a field trip for the learners to one of the many sites across South Africa where fossils can be seen both in museums and *in situ*. It may also be combined with a unit or lessons on dinosaurs which are fascinating to young children. Geological time frames can be introduced.

These activities would lead them through the scientific method where they would:

- observe phenomena;
- formulate questions;
- observe changes;
- record data; and
- explain in words or writing what they have observed.[1]

What is learned

In Grade 1 learners will not fully understand the idea of millions of years, but extreme numbers and sizes, large and small, are fascinating to them. Many of the provinces have some fossil sites and/or museums with fossil displays. The fossil history of South Africa is a crucial link in understanding the history of human evolution; *Sterkfontein* is a World Heritage Site because of its importance in this regard. Grade 1 provides a valuable opportunity to start exposing our learners to the notion of change over time in ways they will enjoy and understand, and that will teach them some science and history as well.

Intermediate Phase textbook unit

Adaptation of organisms can be dealt with in the Intermediate phase. A textbook unit would provide activities through which learners begin to make sense of how organisms live in their environments, and to develop an awareness of the diversity and interactions of living and non-living parts of their environment.

---

1 Department of Education, 1997, *Natural Sciences Policy Documents for Curriculum 2005*, Department of Education: Pretoria.

Activities could include the following:

- Learners observe a specific organism in a local environment, and list as many adaptations as they can observe. They describe the adaptation in terms of function.
- Learners predict and test the response of mealie worms, snails or similar animals to various environmental variables such as light, touch, moisture and temperature.
- The class is divided into two groups of learners who are assigned to separate areas of the school grounds, and are asked to find a specific location in which to hide a camouflaged animal. Learners then make the animals in the classroom (during the art lesson perhaps?) using vegetables such as carrots, potatoes and/or any other materials that they or the educator can access. The learners then put their animals in the chosen locations. The two groups change areas and become 'predators' looking for each other's animals. After a time, they identify which animals they found and it is revealed to them those they did not find. A class discussion is held to identify reasons why some animals were found and some not.

What is learned

To assess learning about the principles of adaptation, the educator could ask the learners to make up questions for each other. Learners could do a written report within a set of given criteria on the experiment using snails or other animals. They could prepare a presentation on their animal, what features the animal uses and how the features allow the animal to adapt well to its environment.

The principles of adaptation are then extrapolated so that learners can use them to explain the variation and change of organisms over time or evolution, as they provide an explanation of why such change might take place.

Senior Phase textbook unit

In the Senior Phase it will be necessary to deal with topics that may provoke confrontations with learners who feel that their religious beliefs are being threatened. Learners will hopefully – by this level

– be familiar with scientific method, geological time, fossil records and the concept of change over time. Classroom educators will have to ask learners to attempt to follow the logic of the presentation in a textbook unit. What follows is an example of how Charles Darwin's theory of natural selection can be presented in a textbook unit:

- Begin with the story of Darwin's carefully recorded observations of 13 different finch species on the Galapagos Islands. Describe the species in terms of their beak shape and size related to the food supply, and describe the mainland finch species in the same way. An activity would be to illustrate some or all of the beak types, along with a variety of foods, and ask learners to observe the illustrations and explain why particular beaks would be better for eating particular foods.
- Discuss artificial selection, which has been used over many hundreds of years by farmers to obtain more productive plant species, and by dog breeders to produce different species of dogs for different purposes.
- Relate this kind of artificial selection to the different species of finch.
- Explain how Darwin's theory of natural selection developed from his observations of the finches, as well as from his own experiments of artificial selection amongst his pigeons.

What is learned

The beak activity requires development of visual literacy, observation and the ability to relate biological structures to their functions. This is important in understanding both how living things work and how evolution or change over time of structures may have come about.

An activity on the development of the theory could provide the information as a written passage which is used as a comprehension in order to develop literacy in general and scientific literacy in particular. Questions are asked about the passage, structured so that learners start with selecting information provided in the passage, and later move on to questions that require learners to explain Darwin's thinking in their own words.

Possible conflict for learners

Learners can be asked to discuss or write how they respond to this topic, and what they feel about what they are learning. This leaves opportunity for learners to express discomfort or disbelief, but also allows them to be exposed to a scientific thinking process.

Science and religion

As well as current curricula, South Africa now has an exceptional constitution, of which an important feature is tolerance of diversity of views, values and speech. Curriculum 2005 and the Revised Curriculum Statements incorporate the values of tolerance that the constitution expresses, in the context of a secular, although diverse, state.

The burning question, is how to present topics related to evolution, that may potentially offend the religious views of some, as a necessary part of the curriculum in the learning areas of the natural sciences and human and social sciences. How can this be resolved while maintaining the principle of valuing diversity? Stephen Jay Gould offers his views on the matter: '...they [Jehovah's Witnesses] do not try to impose their theological beliefs upon public school science curricula, and they agree with my view that churches and homes are the proper venues for teaching such private and partisan doctrines.'[2]

Gould also writes good advice to a natural or social science educator: 'I also do not understand why the two enterprises (science and religion) should experience any conflict. Science tries to document the factual character or the natural world, and to develop theories that co-ordinate and explain these facts. Religion, by contrast, operates in the equally important, but utterly different, realm of human purposes, meanings and values – subjects that the factual domain of science might illuminate, but can never resolve. Similarly, while scientists must operate with ethical principles, some specific to their practice, the validity of these principles can never be inferred from the factual discoveries of science.'[3]

2 Gould, S.J. 1999, *Rocks of Ages: Science and Religion in the Fullness of Life*, Vintage: London; Ballantine: New York, p.127.

3 Gould, S.J. 1999, *Rocks of Ages*, p.4.

This passage is quoted at length to suggest that the job of school Science educators is to present a framework of investigations and information within which learners can comprehend how a theory to explain observable facts may have arisen. There is no adequate solution to the conflict between the religious view of creation and the scientific theory of evolution except one of tolerance and engagement.

Conclusion

The examples provided above illustrate ways of presenting scientific learning related to the topic of evolution in textbooks for schools; they do not provide the answer to the conflict between religious belief and scientific theory. The examples are intended to highlight the challenges provided by this topic, and to help develop a methodology that allows learners to become exposed to a scientific way of thinking which does not conflict with other ways of viewing the world.

The Genome, Biology and Education

4

Science, Genomics and Education in South Africa

Kader Asmal

Credulity is the man's weakness, but the child's strength – Charles Lamb, Essays of Elia.[1]

Science has the nasty, liberating habit of insisting on fact, of testing whether a suspicion or assumption is true or not, of taking truth as provisional until proven otherwise. That is how it should be; and that is why science has made such staggering cumulative progress, bringing us to the point of the Human Genome Project, the extraordinary intellectual and technological product whereby we are able to reveal the genetic basis of our being, and to begin now to make sense of it all. The promise of meaning for advancing human welfare, in history, archaeology, paleo-anthropology, health, medicine, biotechnology, ethics and law are staggering, to put it mildly. We are on the cusp of rewriting world history and advancing human welfare as never before.

Science, though, has no built-in moral compass, no natural leaning towards doing what is right and ethically appropriate.

1 Lamb, C. 1993, *Essays of Elia*, Carcanet Press [originally published in 1895, H Ahenus: London].

Although science has some fascinating things to say about the genetic basis of our ability to be moral and ethical – and in particular the evolution of the human brain and its neurology and neuro-psychology, a flowering field of scholarly enquiry today – it has no inherent philosophical foundations for making moral claims about the world.

This is why I admire the distinction made by the late Stephen Jay Gould, in his *Rocks of Ages: Science and Religion in the Fullness of Life*, between the magisteria of science and of religion.[2] Gould deserves greater than a parenthetical tribute to his contribution to evolutionary biology; his extraordinary ability to bring together disparate observations into one line of argument; his charming literary ability; and his persistent efforts in bringing science into the public domain, a modesty of inclination rarely seen in the mighty and sometimes aloof and distant world of science scholarship. Pippa Skotnes, of the Michaelis School of Fine Art at the University of Cape Town, for whose wonderful book *Sound from the Thinking Strings* he wrote the introduction,[3] gave an appropriate tribute to Gould's life and work in this book.

We do not live enough by science, that much is true; but we do not and cannot live by science alone, that much is also true. Thus the sub-title of Gould's book, which I see was borrowed for the title of the Colloquium, 'Science and Religion in the Fullness of Life', must be correct in recognising the complexity of our being and consciousness. His plea for an honest and rational engagement between science and religion – and may I add that, by including Catherine Odora Hoppers in this book, the wisdom of indigenous knowledge – and their related knowledge systems is also valid. Everyone would surely agree that our lives would be fuller if science and religion were to co-exist in their active embrace for the truth by engaging in robust and courageous dialogue.

In that engagement, science will unravel and discredit some, perhaps even many, of the beliefs and suppositions of certain traditions of faith and religion. For most people, these are matters

2 Gould, S.J. 2002, *Rocks of Ages: Science and Religion in the Fullness of Life*, Vintage: London; Ballantine: New York.

3 Skotnes, P. (ed.) 1991, *Sound from the Thinking Strings. Foreword* by Stephen J Gould. Cape Town.

requiring reflection and resolution, and if various papal authorities can live by evolution then surely so can we. It is the dogmatists of religion who still insist that we are but 5 000 or so years old when the evidence overwhelmingly points towards a much, much longer ancestry; that we are not ready-made, perfect physiological constructions, immutable to evolutionary change. Divine creations we may ultimately turn out to be, for scientists would hesitate to declare on the early origins of living matter, but we are certainly the results, the consequences, of evolution.

I believe that there are publishers who fear producing texts on evolution because of our supposedly official ambivalence on this matter; let me say now that there is no ambivalence, as our revised curriculum statements on the Natural Sciences affirm without a doubt that evolution is a central part of what is to be taught in the Natural Sciences, and we need textbooks as necessary aids to its teaching. Please, let books be produced in substantial numbers. We need them. Our teachers need them. Our learners in school need them.

In the Department of Education, we uphold the distinction between religious education and religion *in* education. As publicly funded schools by the grace of taxes, we believe that it is proper and desirable to teach all of our children in schools about the rich history of our religious diversity.[4] Our children benefit, we believe, if we share our admiration and understanding with them of all religions and faiths – Christianity, Islam, Judaism, Hinduism and indigenous faith traditions – for in so doing we affirm our common humanity. The teaching of religion as faith history is what ought to happen in our public schools.

We advocate this also because we believe an approach of this kind promotes tolerance. Religion, of course, has been the source of extraordinary intolerance, but it has also been the inspiration *for* tolerance, in the deep sense of providing the basis for what the Working Group in Values in Education describes as reciprocal altruism and an 'active appreciation of the value of human

4 See Chidester, D. 2002, 'Religion Education, Learning about Religion, Religions and Religious Diversity', in Asmal, K. & James, W. (eds.) *Spirit of the Nation: Reflections on South Africa's Educational Ethos,* New Africa Education: Cape Town.

difference'.[5] Indeed, the *Manifesto on Values, Education and Democracy*, published in 2002, recognises the positive role of certain faith traditions in forging our democratic transformation and in nurturing tolerance in our multi-religious and multi-cultural society.[6]

I mention this in order to affirm Gould's notion that 'science' and 'religion' are central building blocks in the 'fullness of life', for one without the other leaves us truncated, incomplete, even bereft of the creations of our past history. From this point of view, we in the Department of Education are committed to the teaching of the best of science and the best of our religious variation in our schools.

Teaching the best of science is, of course, a challenge. Not least as a result of the awful and criminal legacy of apartheid – that used racist anthropology to justify setting science and mathematics aside for white, coloured and Indian South Africans, thus robbing Africans of decades of science learning. Not that science and mathematics teaching for whites, coloureds and Indians was equally endowed – it was not – nor that they did not also suffer miseducation on a wide variety of fronts. The idea, though, that the large majority of our population was not fit or desirable beneficiaries of science in the century of science for decades of our history, is the cowardly and pernicious legacy of the brutally dishonest misreading of the biblical injunction that South Africans bearing merely a darker skin were the 'hewers of wood and drawers of water.'

My Deputy Minister Mosibudi Mangena is energetically working to overcome the scientific backlog that previous educational policies allowed to accumulate in our society. We are of course continually updating our syllabi, and the recently released Revised Curriculum Statement for Grades R–9 in the Natural Sciences, goes a long way to bringing the best of the latest science to our younger learners. The Department has further initiated two critical mathematics, science and technology projects:

5 Working Group on Values in Education, *Values, Education and Democracy* (Cape Town, 2000).

6 James, W. (ed.) *Manifesto on Values, Education and Democracy*, Cape Argus Teach Fund: Cape Town.

(1) A reskilling and upgrading project for teachers in the Intermediate and Senior Phases; and
(2) The *Dinaledi*, a Mathematics, Science and Technology project that reaches out to 102 senior secondary schools.

'The two projects are part of a rollout of the strategy we launched in June 2001, a strategy whose motto is *Creating Tomorrow's Stars Today*' says the Deputy Minister Mosibudi Mangena in response to my budget speech of Thursday, 6 June 2002.[7]

But the Deputy Minister also tells us that '[O]ur analysis has shown that, whereas the overall participation rate in mathematics and science has remained steady over the years, the number of learners registering for this subject in Higher Grade has been decreasing. This has been accompanied by a marked increase in learners registering for these subjects in Standard Grade. Against all expectations, however, the pass percentage remains lower in Standard Grade than in Higher Grade by an average of between one and three percentage points.'[8]

How can this be? How is it possible for the greatest contributions to making sense of the greatest puzzles of our lives – the issues that should excite the young minds at our schools – fail to stimulate the imagination and inspire by force of interest and enthusiasm better performance? In Jeff Lever's paper, he says that learners experience the teaching of science and biology as 'boring', uninteresting, tedious.[9] Deputy Minister Mangena pleads for a future in which 'we must have millions of young, hot and breathing bodies sitting behind desks in classrooms all over our country successfully learning mathematics, science and technology-related subjects'; out of love for the subjects, he adds.[10]

Mark Shuttleworth and the First African in Space project shows that it is possible to stimulate interest in science, by making it 'cool' and something peer groups admire. The idea of making science cool, though not Shuttleworth's budget that stood behind it, is

7 Mosibudi, M., Deputy Minister of Education, 'National Assembly Debate on the Education Vote' (Cape Town, 6 June 2002) in response to Professor K. Asmal, MP, Minister of Education 'Introducing the Debate on the Education Budget, Vote 14' (Cape Town, 6 June 2002).

8 Mangena, M. 2002, 'National Assembly Debate on the Education Vote', mss p.5.

9 Lever, J. 2002, *Science, Evolution and Schooling in South Africa*, Africa Human Genome Initiative: Cape Town.

10 Mangena, M. 2002, 'National Assembly Debate on the Education Vote', mss p.1.

something, that, if grasped could be nurtured into the love for solving life's puzzles, for taking these wonderful discoveries and adventures of science – things that are certainly not boring or uninteresting or tedious – and turning them into a love of learning, a thirst for knowledge.

But if this is to succeed, we will need the efforts of everyone in the field to ensure that the best materials reach our teachers so that they are equipped to meet the challenges of the classroom. The teacher is key: we know that a passionate, educated teacher will nurture interest in a subject. To teach well requires excellent materials, and high quality, up-to-date reading matter and books. Moreover, *we* rely heavily on the advice and work of many experts in shaping and reshaping our syllabi and our schooling. We shall certainly rely on you all to meet the challenges presented by the developments in human biology and in the aftermath of the completion of the Human Genome Project.

It is heartening to learn, as I have, of the share that South African scientists are taking in the advancement of this field. I know that our universities are promoting numerous projects of both a theoretical and practical nature in relation to this latest and perhaps greatest of recent scientific breakthroughs. Here follows mention of a handful of the ventures that have come to my attention:

- At the University of the Witwatersrand, Dr Himla Soodyall and her colleagues are building on the pioneering efforts of Professor Trefor Jenkins in studying the diverse genetic heritage of our southern African populations. This is one of the most exciting outcomes of the research tools that modern genomics have placed at our disposal. By following imprints left in our DNA by hundreds of thousands of years of development, we can begin to map the movements of the first modern human populations. Dr Soodyall and her team have developed data that amply confirm the antiquity of the indigenous populations of our region and demonstrate long-hypothesised links among various groupings, including the Lemma group and Semitic incomers.
- My own University, that of the Western Cape, has shown considerable foresight in welcoming the initiative by Dr Winston Hide and his collaborators to set up a facility on its

campus for research and training in bio-informatics. The South African National Bio-informatics Institute at UWC is the foremost of its kind in the country, and possibly even the African continent. We know of the immense computational challenges presented by genomic analysis, and its indispensable role in the field. Without trained Bio-informaticians we might as well forget about partaking in the fruits of modern genomic research in general, and human genomics in particular. Particularly pleasing are the intensive short courses that Dr Hide's Institute offers to postgraduates from both South Africa and wider afield.

- At the University of Cape Town, Professor Rajkumar Ramesar, head of the Department of Human Genetics, is utilising the skills of his laboratory and colleagues to help clinicians in their diagnosis and treatment of cancer patients. Not content with sitting in his office or lab, Professor Ramesar also leads a project that travels regularly to the communities of Namaqualand to diagnose and treat families in which there is a history of colon cancer. The project is a model of how the tools of modern genomics can be harnessed and brought to benefit the community itself.

These initiatives are but some of the ventures that prove that South Africa is able to take its place at the forefront of modern molecular biology and the human genome revolution. We constitute a centre of leadership for the rest of the continent but no doubt we could do more: we can do great things with new knowledge of our genes, rewrite African and possibly world history, intervene in the public health sector and develop appropriate biotechnologies – all in service of human welfare.

I note that, in countries often ranked on a similar rung of development as our own, genomics and all its manifold-related fields are being taken seriously indeed. In Brazil, as the recent World Health Organisation report on Genomics and World Health records, strategies that have been adopted over the last five years 'have propelled the country into the upper echelons of international genomics research'.[11] Over 70 laboratories scattered

11 World Health Organisation, 2002, *Genomics and World Health,* WHO: Geneva, mss pp.122 & 124.

throughout this large country are now involved in DNA sequencing work. The government of India established a Department of Biotechnology in 1986. Today, over 4 000 postgraduate students are working in this area.

It is clear that we in South Africa need to emulate these countries. In this regard I would like to congratulate the Human Sciences Research Council, the Academy of Sciences and the Spier Institute for conceiving the Africa Human Genome Initiative. I also wish to say the initiative and its planned conferences cover an extraordinary range of topics, not least of which is the focus on the ethical, legal and educational challenges and risks we face in making the biomedicine and biotechnology of the human and other genomes, assets for our development, in much the same way that the New Partnership for Africa (NEPAD) envisions these to be. The aim of the venture, as I understand, is to promote the development of genomics knowledge and capacity in our own country and fellow African nations. As such, it must complement President Mbeki's ideals for the New Partnership for African Development (NEPAD).

A large and daunting task of the Initiative is to promote public education and awareness of the new scientific era ushered in by the Human Genome. This will be no easy task. To cite the WHO report:

> Genomics has the potential to make an important impact on health care for the future and it is vital therefore that society is prepared for the many new concepts that it will bring with it. This will only be achieved by improving the level of education about the basics of genetics at every level, ranging from primary schools to adult populations. Because of the time scale that will be required to achieve this end, and because of the speed of development and discovery in genomics, it is important that this process is started without delay.[12]

Our new school syllabi will go some way towards providing the basis for this social awareness in the younger generation, but we will need to reach out beyond the schools to the general public. This goal will require the combined efforts of experts, the media,

12 World Health Organisation, 2002, *Genomics and World Health*, mss pp.122 & 124.

educationalists and government departments if it is to be at all effective.

Here the Africa Human Genome Initiative can play a major role in helping co-ordinate and promote public awareness about the opportunities for improved health and for the expansion of our economy that the genomics revolution offers to those who are receptive to its fruits.

It goes without saying that the new frontier opened by genomics has other, far-reaching, implications. As a lawyer and politician, I am only too aware of the intricacies of intellectual property rights that arise in this field. We need to ensure that there are experts in our Law faculties who have their fingers on the pulse of this issue. We must reap the benefits of our work, and prevent the abuse of our own biological resources by outsiders more interested in profit than social welfare. Whatever the developments the field of genomics brings forth, we must be prepared to deal with them, in ethical as well as legal terms. Our religious communities too must be consulted and kept abreast of the imminent changes, so they can offer their members informed guidance.

I have referred mainly to human biology, but of course the findings of genomic analysis encompass all forms of living matter. Plant genomics is a rapidly growing field that wisely applied may generate untold benefits in agriculture, horticulture and of course pharmaceutical products. In this regard it is pleasing to learn of the co-operative ventures that our own National Botanical Institute, with its eight-member Botanic Gardens, is undertaking with such leading scientific institutions as the Royal Botanical Gardens in Kew, London. The benefits for African development envisaged by NEPAD can be well served by such collaborations.

I am sure that there are many other exciting developments taking place. The Africa Human Genome Initiative will be undertaking an audit of our resources in this regard to ensure that our scientists and their co-workers receive the credit and public acclaim that is their due. We are fortunate to have a community of scientists equipped to engage and build upon the latest and best of international science. For our part, let us support and expand their efforts in whatever way we are called upon to do.

This book is the beginning of such an endeavour. The colloquium on which it is based is a wonderful illustration of the South African possibility of rational dialogue between individuals from diverse backgrounds and indeed adds to the 'fullness of life'. I look forward to hearing more from the Africa Human Genome Initiative, for we must, in the haunting words of the African-American poet Melvin B Tolson, advance together:

> 'Out of abysses of illiteracy,
> Through labyrinths of lies,
> Across waste lands of disease …
> We advance.' [13]

---

13 From: Tolson, M.B. 1944, 'Dark Symphony' in *Rendezvous with America*, Dodd, Mead & Company: New York.

About the Authors

Professor Kader Asmal, MP, is Minister of Education in the South African Government. Between 1994 and 1999 he was Minister of Water Affairs and Forestry. He has written or co-authored five books – most notably *Reconciliation through Truth* – and co-edited two others. He has been awarded honorary degrees from a number of universities and is an Honorary Fellow of the London School of Economics. He is also an Honorary Professor of the University of the Western Cape.

Professor David Chidester is a Visiting Fellow in the Social Cohesion and Integration Research Programme. He is Professor of Comparative Religion at the University of Cape Town, Director of the Institute for Comparative Religion in Southern Africa (ICRSA), and Co-Director of the International Human Rights Exchange. He is author or editor of fifteen books in the study of religion, including *Religions of South Africa (1992), Shots in the Streets: Violence and Religion in South Africa (1992), Savage Systems: Colonialism and Comparative Religion in Southern Africa (1996), and Christianity: A Global History (2000).*

Professor Linda Chisholm is a director of the Human Sciences Research Council's Education Policy Research division. She has written and published widely on all aspects of education and is a member of various local and international professional and research bodies that aim to improve schools. She chaired the Ministerial Review Committee on Curriculum 2005 whose report made specific recommendations as to how the national schools curriculum for Grades R–9 should be improved, as well as the Ministerial Project Committee appointed to produce the Revised National Curriculum Statement.

Ms Fathima Dada is currently on two Ministerial Project Committees, one to develop the General Education and Training Curriculum and the other to develop the Further Education and Training Curriculum. She is the Chairperson of the Board of Johnnic Learning and Chairperson of Pearson Education Africa. She was formerly the Chief Executive Officer of Maskew Miller Longman Educational Publishing.

She is widely published on the subject of education and has also published several children's books. Her other interests lie in the relationship between the media and the delivery of education.

Judge Denis Davis was appointed a Judge of the High Court in October 1998. He was technical legal advisor to the Multi-party Negotiating Forum and Constitutional Assembly, which were the two fora responsible for drafting the final Constitution. Best known for his role as a television presenter on '*Future Imperfect*' and '*Constitutional Court*', he has also authored or co-authored eight books on subjects as diverse as criminology, estate planning and constitutional law.

Ms Colleen Dawson has been involved in different aspects of science education and material development for over twenty years, as a teacher, publisher, curriculum developer and author. She has constructed teaching curricula for learners in all grades to help them understand evolutionary theory without compromising their religious or ethical beliefs. Most recently she has developed these ideas in the writing of various textbooks.

Professor Wieland Gevers is President of the Academy of Science of South Africa and Chairperson of the Education Committee of the South African Universities' Vice-Chancellors Association. He has been Senior Deputy Vice-Chancellor responsible for planning and academic process at the University of Cape Town since 1992 and Professor of Medical Biochemistry since 1978. He directed Medical Research Council (MRC) Research Units at both the Universities of Stellenbosch and Cape Town and has been awarded the Wellcome Gold Medal and the Gold Medals of both the South African Society for Biochemistry and Molecular Biology, and the South African MRC.

Dr Thavendran (Dev) Isaac is the National Co-ordinator for the Natural Sciences Working Group, tasked with the responsibility of streamlining and strengthening Curriculum 2005 and producing the Learning Area Statement for Natural Sciences. He is currently the Chairperson of both the South African and KwaZulu-Natal Associations of Science and Technology Educators. He is widely published and was recently honoured by the National Science and Technology Forum for having made a most significant contribution in the science, engineering and technology fields.

Dr Wilmot James is an executive director of the Human Sciences Research Council, responsible for the directorate Social Cohesion and Integration Research. He is the author or editor or co-editor of 11 books and monographs, the most recent being *Spirit of the Nation: Reflections on South Africa's Educational Ethos,* co-edited with Kader Asmal. He is formerly the Associate Editor of the *Cape Argus,* Dean of Humanities at the University of Cape Town and Executive Director of the Institute for Democracy in South Africa. He presently serves on the Electoral Task Team for the Department of Home Affairs.

Professor Bernard Lategan is Director of the Stellenbosch Institute for Advanced Studies. He was formerly Dean of Humanities & Social Sciences and served ten years as SAFCERT Moderator for Biblical Studies. He developed programmes in Value and Policy Studies and also teaches value studies to students in commerce and business. He often facilitates strategic processes in organisations and companies aimed at the development of common values.

Dr Jeffrey Lever is a consultant to the Africa Human Genome Initiative and a visiting fellow of the Department of Sociology, Rhodes University. He was editor of the South African Sociological Review and is joint managing editor of the African Sociological Review. He is the co-author of 'The Second Republic: Race, Inequality & Democracy in South Africa' in C.V. Hamilton, et. al., (eds.), *Beyond Racism: Race & Inequality in Brazil, South Africa and the United States.*

Professor Catherine Odora Hoppers is an Associate Professor of the Faculty of Education of the University of Pretoria and Coordinator (Corporate) of the project on *Integration of Knowledge Systems* (IKS-South Africa). She was formerly a Chief Research Specialist for the Human Sciences Research Council and Deputy Director of the Centre for Education Policy Development, Evaluation and Management. She is widely published and has addressed several national and international conferences on fields such as social policy and development, education, gender and epistemology.

Ms Naledi Pandor, MP, is the Chairperson of the National Council of Provinces. In 2000, she was elected the first woman Chancellor of the Cape Technikon and was formerly the Deputy Chief Whip of the African National Congress. She also served in several portfolio committees, most notably as convenor of the sub-committee on higher education in the Education Portfolio Committee.

Professor Pippa Skotnes is Director of the Michaelis School of Fine Art at the University of Cape Town. Her research interests include museums, rock art, the politics of representation and reception as well as Catholic ideas of sacrifice, death and resurrection.

Professor Mark Swilling is Academic Director of the Sustainability Institute, a non-profit trust that works in partnership with the University of Stellenbosch on education and research related to sustainable development. He is also Extraordinary Professor in the School of Public Management and Planning at the University of Stellenbosch, and a Senior Research Fellow at the Warwick Institute of Governance and Public Management, Warwick University (UK). He is widely published and is on the Editorial Board of leading academic journals, namely *Urban Forum, Administratio Publico,* and *Politeia.*

Professor Abdulkader Tayob has been teaching Religious Studies and Islamic Studies at the University of Cape Town since 1989. He has published four books on the Islamic religion. His main focus has been to try and make sense of Muslim discourse in the context of apartheid and post-apartheid South Africa. His other interest is the emergence of Islamic thought and disciplines in the first three centuries of Islam.

Ms Lynne Wilson is Senior Editorial Specialist in the Social Cohesion and Integration Programme of the HSRC. She has spent the past 24 years working as a writer or sub editor on a number of prominent South African newspapers, as well as at *The Independent, The Guardian* and the *Financial Times* in London.

Index

T

U

V

W